Electronics for technicia

Electronics for technicians

R. H. Joynson, B. Eng., C. Eng., M.I.E.R.E., M.I.E.E.E.
Head of the Department of Electrical and Electronic Engineering,
North West Kent College of Technology

Edward Arnold

First published in Great Britain 1983
by Edward Arnold (Publishers) Ltd
41 Bedford Square
London WC1B 3DQ

Edward Arnold (Australia) Pty Ltd
80 Waverley Road
Caulfield East 3145
PO Box 234
Melbourne

British Library Cataloguing in Publication Data

Joynson, R. H.
Electronics for technicians.
1. Electronics
I. Title
537.5 TK7815

ISBN 0-7131-3484-4

Text set in 10/11pt Compugraphic English Times by Colset Private Ltd, Singapore.
Printed in Great Britain by Richard Clay (The Chaucer Press) Ltd, Bungay, Suffolk.

Contents

Preface

This book is intended primarily for students studying for a certificate or diploma awarded by the Technician Education Council and covers the electronics objectives at levels 2 and 3. I hope that it will also be of use as general text for those who are following other syllabuses or who simply wish to gain a further understanding of electronics. It is assumed that readers will have a knowledge of or be studying basic electrical principles.

Once again I would like to acknowledge my debt to all those sources of information which, as a teacher, one invariably consults and whose source is often forgotten as time passes. Particularly worthy of mention are the data books and technical manuals produced by the various manufacturers.

I wish to offer my thanks to my wife, Kath, for her tolerance and understanding during those times when she thought I was on holiday. I would particularly like to thank my colleagues Tom Murray for his assistance and comments during the preparation of the manuscript and Rob Campbell for checking the practical exercises for me. Finally, I would like to thank Bob Davenport of Edward Arnold (Publishers) Ltd for his usual thorough reading of the manuscript at all stages, which has resulted in a much better final offering than would otherwise be the case.

Roy Joynson

Acknowledgements

The author wishes to thank the following organisations for kind permission to reproduce photographs of their equipment: Mullard Ltd (figs 11.1, 11.2, and 11.4) and A. M. Lock & Co. Ltd (fig. A1.1).

Thanks are also due to Mullard Ltd for permission to make use of copyright material and drawings as follows: fig. 11.3, the explanation of integrated-circuit manufacture, Tables 11.1 to 11.4, and data sheets for the following devices: BC107–109, BFW10, NE/SE5539, μA741, LM111, CA3089, and TDA1072.

The author would also like to thank A. M. Lock & Co. Ltd for providing the Locktronics LK750 baseboard and components used for building the practical circuits.

1 Semiconductor diodes

1.1 Introduction

The 'diode' is the simplest practical device of the semiconductor family. One of the most important uses for the diode is to convert alternating current into direct current in power supplies. Diodes are also widely used in electronics to carry out a variety of functions such as clipping, clamping, protection, and detection. Special diodes such as the zener diode are used in stabilising and protection circuits.

The structure of semiconductors will be considered using a very simple and stylised model, since we are primarily concerned with the applications rather than with the physics of the devices.

1.2 Simplified atomic theory

For practical purposes an atom can be considered as a solid consisting of the nucleus plus the inner electrons surrounded by the valence or outer-shell electrons. The model for semiconductors is shown in fig. 1.1; other elements will have different numbers of valence electrons. A full outer shell usually has eight electrons in it.

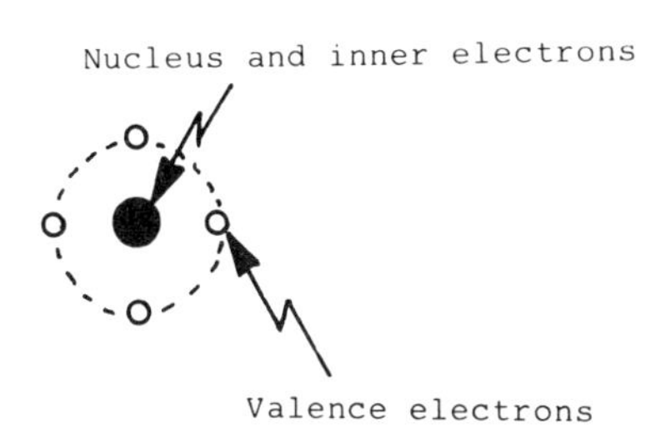

Fig. 1.1 Model semiconductor atom

The most important materials in modern electronics are made up of silicon or germanium. These elements have a valency of four ('valency' refers to the number of electrons in the outer shell or layer). Their atoms form a cubic crystal lattice in which the orbits of the outer electrons overlap, resulting in each atom 'sharing' electrons with each of its four neighbouring atoms. This form of sharing – known as covalent bonding – establishes a very stable configuration, since each atom is 'fooled' into thinking it has a full outer shell and is thus effectively inert. This is shown in fig. 1.2.

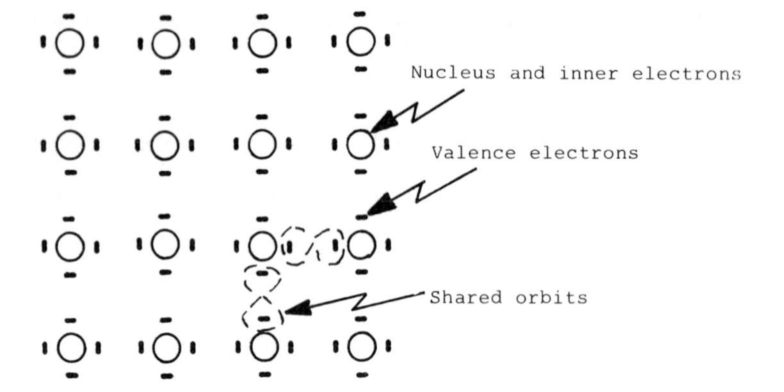

Fig. 1.2 Covalent bonding. (For clarity, only two sets of shared orbits are shown.)

1.3 Energy levels in solids

According to the generally accepted view of quantum theory, the electrons orbiting around the nucleus of an atom can only have certain discrete or fixed energies. These electrons inhabit fixed 'energy bands' or 'levels'. Figure 1.3 illustrates a simplified energy-band diagram. The lower energy levels, below the valence band, are completely full, so no movement of electrons and hence no current flow is possible in these bands. The bands of most interest in semiconductors are the valence band, which is filled with electrons, and the conduction band which is empty. If sufficient energy, usually in the form of heat, is given to the material, electrons may move into the emission band before being emitted from the surface.

The spaces between the bands are known as 'forbidden bands' or 'gaps', and an electron cannot occupy a 'forbidden' energy level. To cross a

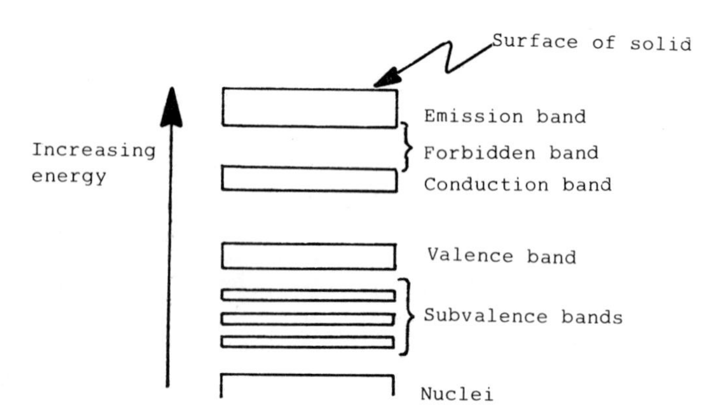

Fig. 1.3 Simplified energy-band diagram for a solid

forbidden band to reach another energy level, an electron must receive or give up a fixed amount of energy before moving to the new level.

1.4 Semiconductors

Semiconductors are materials whose electrical properties fall between those of the insulators and the conductors. The resistivities of some of the common conductors, insulators, and semiconductors are listed in Table 1.1.

Table 1.1 Resistivities of common conductors, insulators, and semiconductors

Material	Typical resistivity (Ω m) at room temperature
Aluminium	2.7×10^{-8}
Copper	1.8×10^{-8}
Silver	1.6×10^{-8}
Silicon	600 (intrinsic value)
Germanium	0.6 (intrinsic value)
Glass	10^{+4}
P.T.F.E.	10^{+12}

A perfect semiconductor can be considered to be a material with all its valence-band electrons fixed and unable to move, because there are no available spaces or 'energy levels' in the valence band into which an electron can move, and an empty conduction band. When a potential is applied to the material, no movement of electrons – and hence no current flow – can take place since before an electron can move there must be an available 'empty' energy level within the valence band into which it can move. Provided the energy supplied is not sufficient to allow an electron to cross the forbidden band, the material is a good insulator.

In practice, at normal room temperatures silicon and germanium have sufficient thermal energy – energy due to the ambient temperature – to allow a few electrons to 'escape' to the conduction band. If a potential is now applied to the material, any of the electrons which have escaped are able to move easily in the conduction band. Also, any electron which moves into the conduction band leaves behind a hole which allows some electrons to move within the valence band. Increasing the temperature allows more electrons to escape to the conduction band; hence the resistance of semiconductors decreases as the temperature is increased, i.e. they have a negative temperature coefficient of resistance. This is the opposite to the effect of temperature on conductors and insulators. Figure 1.4 shows typical resistance/temperature characteristics for some common conductors, insulators, and semiconductors.

The properties of the most common semiconductor materials can be modified by the addition of small quantities (from 1 part in 10^{10} to 1 part in 10^{6}, depending on the type of device) of suitable impurity elements with different numbers of valence electrons per atom – a process known as

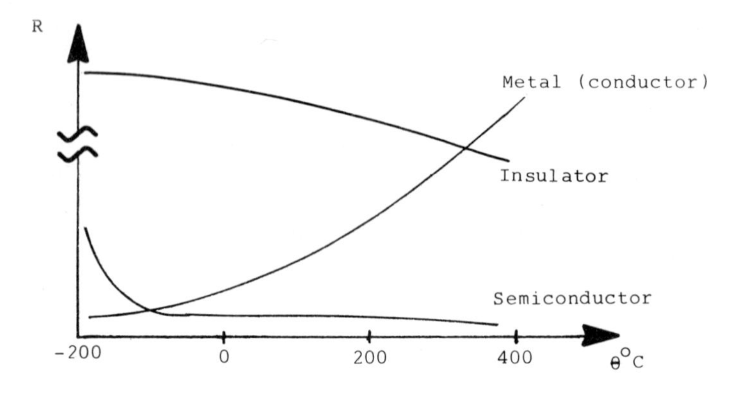

Fig. 1.4 Typical resistance/temperature characteristics

'doping'. If a few pentavalent atoms – i.e. atoms with five valence electrons, e.g. antimony or arsenic – are added to the silicon (or germanium), each of these atoms will make available an electron which can easily move into the conduction band and carry current, fig. 1.5. Since electrons are negatively charged particles, this doped material is known as 'n-type' semiconductor material. The pentavalent atoms which have each 'given' an electron are known as 'donor' atoms.

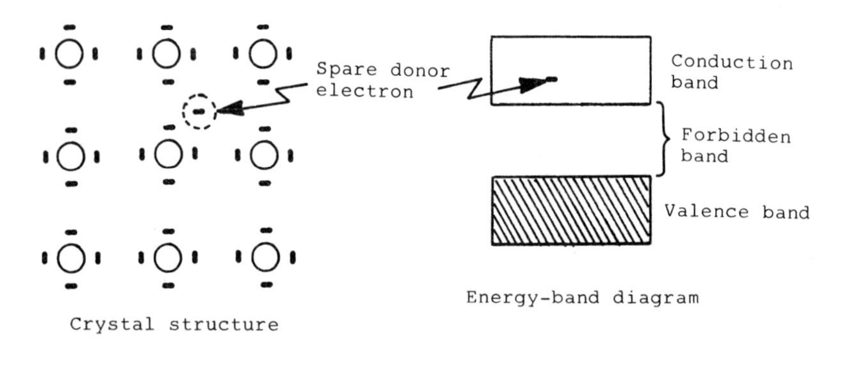

Fig. 1.5 Effect of pentavalent doping

If a few trivalent atoms – i.e. atoms with three valence electrons, e.g. indium or boron – are added to the silicon (or germanium), a space or 'hole' is created in the lattice, fig. 1.6. This 'hole' in the normally full valence band effectively acts like a positive charge moving in the opposite direction to an electron, even though its effect is in fact due to electrons moving into the vacant holes in the lattice. Certain phenomena cannot be explained without the concept of a positive charge such as the 'hole'. The holes behave as positive charge carriers, and this material is referred to as

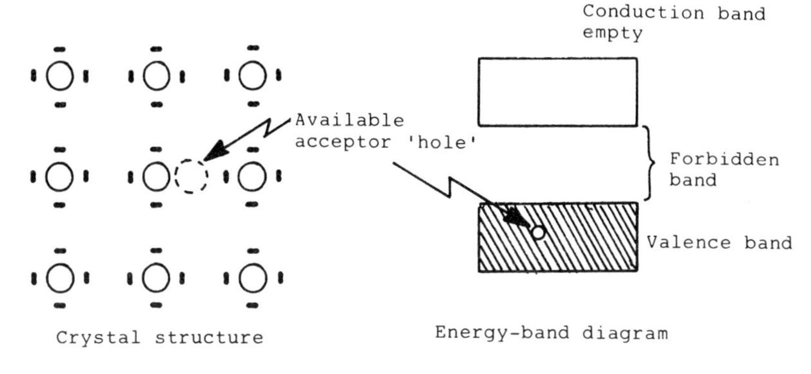

Fig. 1.6 Effect of trivalent doping

'p-type' material. The trivalent atoms which are each able to receive an extra electron are known as 'acceptor' atoms.

The electrons in n-type material and the holes in p-type material are *majority* carriers. At room temperature, a small number of thermally generated holes in n-type material and electrons in p-type material are present and these are known as *minority* carriers. Normally these minority carriers can be ignored; however, increasing temperature generates more minority carriers and eventually results in the loss of any special semiconductor properties. Silicon devices are able to operate at a higher temperature than germanium before these properties are lost, about 200°C compared to 75°C.

Pure semiconductors are referred to as *intrinsic* and doped semiconductors as *extrinsic* semiconductors.

1.5 Junction rectifiers

The principle of operation of a junction rectifier is the same for silicon and germanium. The junction consists of one piece of n-type and one piece of p-type material formed in the same crystal lattice.

The effects of forming the junction are shown in fig. 1.7. Electrons from the n-type material cross the junction to the p-type material and fill

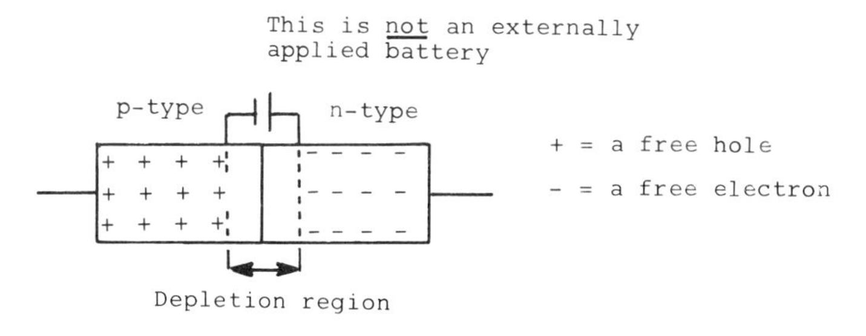

Fig. 1.7 p–n junction after formation

some holes, while some holes cross the junction in the opposite direction and 'collect' electrons. This process is known as *diffusion*. As a result of this diffusion, a small area on the 'n' side of the junction becomes positively charged and a small area on the 'p' side becomes negatively charged; i.e. the 'n-type' material is less negative and the 'p-type' material is less positive. The effect of these charges is to set up an equilibrium condition which prevents further movement of electrons or holes across the junction, since the negative charge in the p-type material repels further electrons and the positive charge in the n-type repels further holes. Because there are no current carriers – holes or electrons – present in this small region on either side of the junction, it is called a depletion region (or layer). Other names used are space-charge region and barrier region, and it is effectively a potential barrier.

1.5.1 Reverse bias

If a voltage is applied to a junction as shown in fig. 1.8, the positive plate of the battery will 'pull' electrons and the negative plate will 'pull' holes away from the junction. This widens the depletion region and, since the depletion region has no charge carriers available, no current will flow. The junction is then said to be *reverse-biased*. In practice, at room temperature a small number of minority carriers are present in each region and these will flow across the junction and form a small 'leakage current'. Provided the device is operated within its specification, this leakage current can be ignored for most practical purposes.

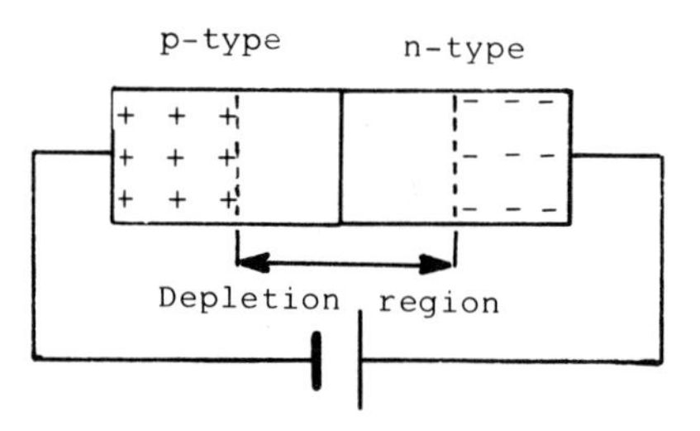

Fig. 1.8 p–n junction with reverse bias (no current flow)

1.5.2 Forward bias

If a voltage is now applied to the junction as shown in fig. 1.9, the positive plate will 'push' holes and the negative plate will 'push' electrons towards the junction. Once the holes and electrons cross the junction, current will flow easily and the device is said to be *forward-biased*.

Typical characteristics (graphs of current against voltage) for the forward and reverse directions for typical silicon and germanium *diodes*, as these

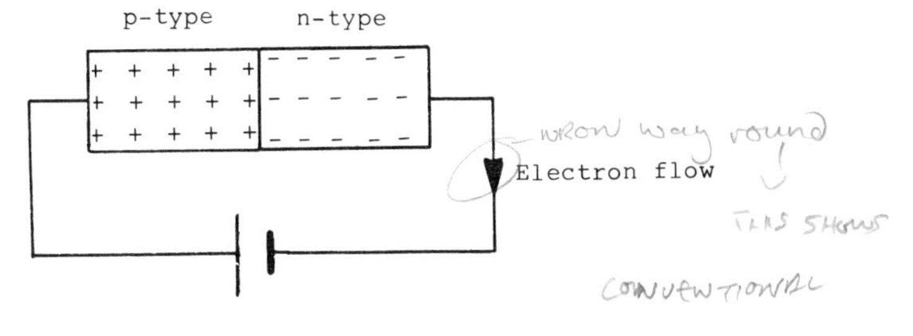

Fig. 1.9 p–n junction with forward bias

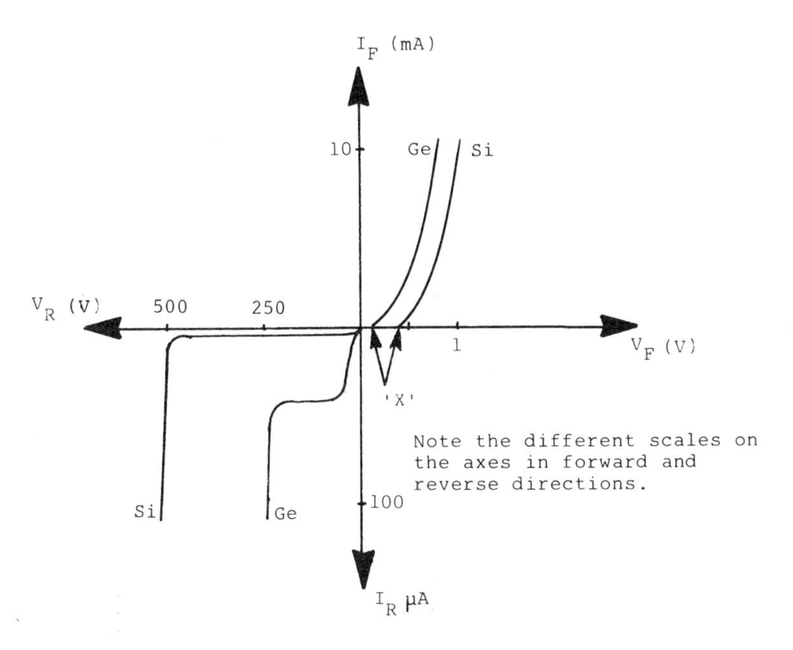

Fig. 1.10 Diode characteristics

devices are known, are shown in fig. 1.10. The potentials at points 'X' on the graph correspond to the potential required to push the holes and electrons across the junction. Typical values for these potentials are 0.2 V for germanium and 0.6 V for silicon.

If the maximum reverse voltage a diode can withstand – the peak inverse voltage (PIV) – is exceeded, the device will be destroyed by the large current which will flow. This large current is caused by the small number of minority carriers being accelerated by the large electric field, giving them sufficient kinetic energy to dislodge other electrons from the lattice if they collide with them (this is a cumulative or avalanche effect). Below breakdown, reverse current is very small compared to the forward current but doubles for every 8°C temperature rise in the case of silicon (every 10°C for germanium).

With all semiconductor devices, high operating temperatures should be avoided since these can cause the crystal to melt. When conducting, the diode is generating heat – for example, a silicon diode carrying 30 A with a forward volt drop of about 0.6 V would dissipate 30 A × 0.6 V = 18 W – thus it may be necessary to use heat sinks to remove excess heat quickly. If the reverse breakdown voltage (PIV) is exceeded, reverse current increases so rapidly that the device melts the crystal, destroying its properties. Silicon rectifiers are made with ratings up to at least 500 A, with PIV's of 3000 V. Germanium devices are not normally rated so high.

Figure 1.11 shows the block diagram, symbol, and actual diode.

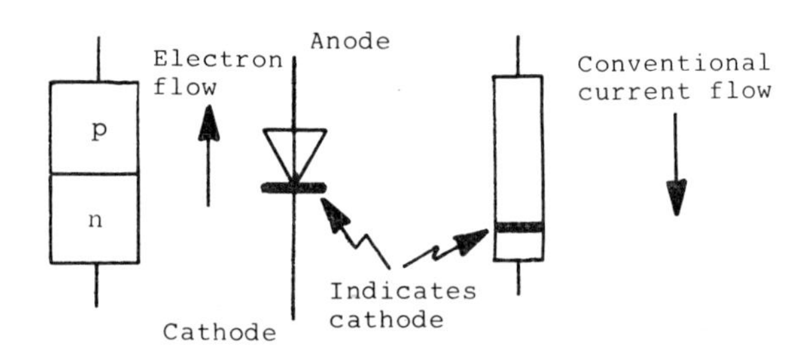

Fig. 1.11 Diode block diagram, symbol, and marking

1.6 The zener diode

In a p–n junction, the depletion region has only a few minority current carriers and if it is reverse-biased – which is normally the case for zener diodes – then it is effectively a very high resistance. If a large voltage is applied, the diode will break down, i.e. conduct. There are two mechanisms causing this breakdown:

i) *Avalanche breakdown* The minority carriers can be accelerated to high velocities by the applied reverse field which gives them sufficient energy to dislodge other electrons from the crystal lattice in the event of a collision, resulting in a cumulative effect. This is the normal mechanism causing reverse breakdown in an ordinary diode.
ii) *The zener effect* Some devices are made with very narrow depletion regions. The reverse bias then gives rise to a strong electric field which is sufficient to cause some of the covalent bonds to break.

Devices designed to have a predictable breakdown voltage are called *zener diodes* regardless of which of the above mechanisms is involved, although in fact both are always involved to some extent. The zener effect predominates at lower voltages and produces a sharper 'knee' in the device characteristic. As a result, two low-voltage zeners in series give a more stable output than a single one, e.g. two 4.5 V instead of a single 9 V.

When breakdown occurs, provided current flow does not result in the permissible power-dissipation level for the device being exceeded, no permanent damage is done.

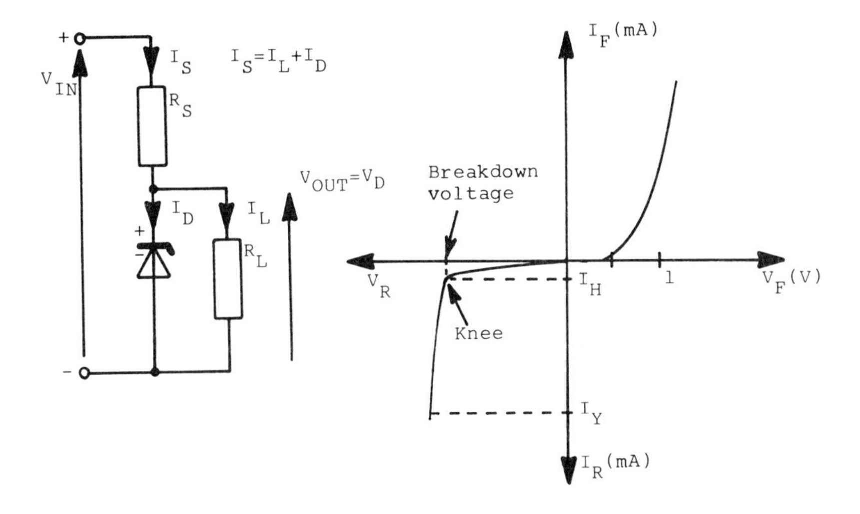

Fig. 1.12 Zener-diode stabiliser circuit and characteristics

Figure 1.12 shows the zener characteristics and a simple application as a voltage regulator or stabiliser. Once the breakdown voltage has been exceeded, the volt drop across the diode remains almost constant even though the reverse current changes from I_H to I_Y. Correct operation can only be achieved when a suitable series resistor, R_S, is used to 'drop' the excess voltage. The zener diode acts as a 'current reservoir' – e.g. if the load requires more current, the diode current will decrease by the required amount to supply it; if the load sheds (gives up) current, the diode current will increase by an equal amount, keeping the current from the supply constant.

The value of the diode current when breakdown occurs is the 'holding current', I_H. In a practical device, this is the minimum current which can be allowed to flow if the device is to function as a stabiliser. For many purposes the device is considered as 'ideal' and the holding current is zero.

Typical calculations involving zener diodes are given in examples 1.1 and 1.2.

Example 1.1 A zener-diode stabiliser circuit (fig. 1.13) is to provide a 12 V stabilised supply to a variable load. The input voltage is constant at 18 V and the diode is a 12 V 400 mW device. Calculate

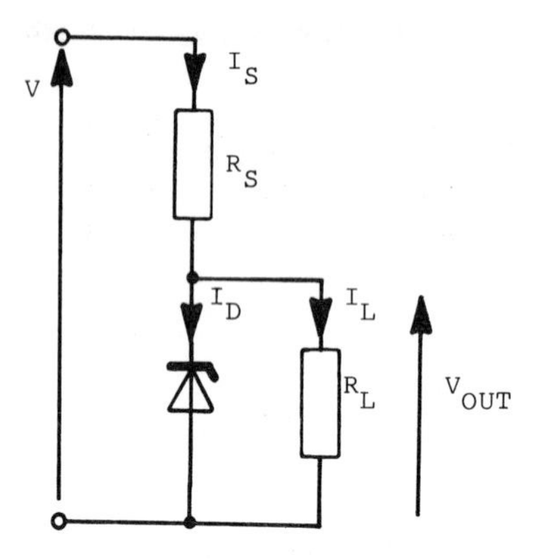

Fig. 1.13 Circuit for examples 1.1 and 1.2

a) a suitable value for the series resistor R_S,
b) the diode current when the load is 2 kΩ.

Figure 1.13 shows the circuit arrangement.

$$I_S = I_D + I_L$$

$$V = V_{RS} + V_D$$

$$P_D = I_D \times V_D$$

a) $$R_S = \frac{V_{RS}}{I_S} = \frac{V - V_D}{I_S}$$

We must now find a value for I_S. If we assume that the diode is ideal, then $I_D = 0$ and the current in the resistor R_S will be the same as the current in the load. However, this is a theoretical case and it is good engineering practice to consider other factors. If the load goes open circuit, then all the current must flow through the zener diode – if the diode cannot handle this, it will be destroyed. Thus a good starting point is the maximum current the zener diode will have to handle. This will occur when $R_L = \infty$, i.e. at open circuit.

$$\therefore \quad I_{DMAX} = \frac{P_D}{V_D} = \frac{400 \text{ mW}}{12 \text{ V}} = 33.3 \text{ mA}$$

$$\therefore \quad R_S = \frac{V - V_D}{I_S} = \frac{18 \text{ V} - 12 \text{ V}}{33.3 \text{ mA}} = 180 \ \Omega$$

As it is virtually impossible to obtain an exact value for all possible resistors, we must use the nearest preferred value (NPV) – in this case it would in fact be 180 Ω. We now have only part of the answer. If you go into a shop and ask for a 180 Ω resistor, the first thing the assistant should ask you is 'What power rating do you want?' So we must calculate a suitable power rating. In this case it is quite simple, since the supply is fixed.

$$\therefore \quad P_{RS} = I_S^2 \times R_S = (33.3 \times 10^{-3}\ \text{A})^2 \times 180\ \Omega = 0.199\ \text{W}$$

Again 199, mW is not a commercially available rating and in any case we would want to allow a safety margin, say a ½ W rating; therefore use a 180 Ω ½ W resistor for R_S.

b) Since V is fixed, I_S must be constant for $R_L = 2\ \text{k}\Omega$.

$$I_L = \frac{V_D}{R_L} = \frac{12\ \text{V}}{2\ \text{k}\Omega} = 6\ \text{mA}$$

and $\quad I_S = 33.3\ \text{mA}$

$$\therefore \quad I_D = I_S - I_L = 33.3\ \text{mA} - 6\ \text{mA} = 27.3\ \text{mA}$$

Example 1.2 A 9.1 V 1.3 W zener diode is to be used as a stabiliser in the circuit shown in fig. 1.13. The supply voltage is 20 V ± 10%, and the load current is constant at 25 mA. Calculate

a) a suitable value for the series resistor R_S,
b) the power dissipated by the zener diode when the supply voltage is maximum,
c) the current in the diode when the supply is at 21 V.

Assume the diode to be ideal.

The circuit diagram is the same as fig. 1.13. The supply voltage varies between 18 V and 22 V.

There are two pôssible methods of solving this problem.

Case 1
a) Consider the situation when the supply voltage is at a minimum:

For $V_{MIN} = 18\ \text{V}, I_D = 0 \quad \therefore \quad I_S = I_L = 25\ \text{mA}$

$$\therefore \quad R_S = \frac{V_{RS}}{I_S} = \frac{18\ \text{V} - 9.1\ \text{V}}{25\ \text{mA}} = 356\ \Omega$$

When the supply voltage is increased to its maximum, 22 V, then I_S will rise to

$$I_{SMAX} = \frac{22\ \text{V} - 9.1\ \text{V}}{R_S} = \frac{22\ \text{V} - 9.1\ \text{V}}{356\ \Omega} = 36.2\ \text{mA}$$

We can now calculate a suitable power rating for R_S:

$$P_{RS} = (I_{SMAX})^2 R_S = (36.2\ \text{mA})^2 \times 356\ \Omega = 466\ \text{mW, say 1 W}$$

b) When V is at a maximum of 22 V, the current I_S in R_S is 36.2 mA (found in part (a) above). The excess current must be flowing in the zener diode, and is equal to

$$I_{SMAX} - I_L = 36.2 \text{ mA} - 25 \text{ mA} = 11.2 \text{ mA}$$

Hence the power dissipated by the zener diode under these conditions is given by

$$P_D = V_D \times I_D = 9.1 \text{ V} \times 11.2 \text{ mA} = 0.1 \text{ W}$$

c) When the supply voltage is 21 V,

$$I_S = \frac{V - V_D}{R_S} = \frac{21 \text{ V} - 9.1 \text{ V}}{356 \ \Omega} = 33.4 \text{ mA}$$

$$\therefore \quad I_D = I_S - I_L = 33.4 \text{ mA} - 25 \text{ mA} = 8.4 \text{ mA}$$

Case 2

We can solve the same problem by considering the position from the maximum-voltage condition. At this point the current in R_S is

$$I_{SMAX} = I_L + I_{DMAX} = 25 \text{ mA} + \frac{P_D}{V_D}$$

where P_D/V_D is the maximum current the zener diode can handle.

$$\therefore \quad I_{SMAX} = 25 \text{ mA} + \frac{1.3 \text{ W}}{9.1 \text{ V}} = 25 \text{ mA} + 143 \text{ mA} = 168 \text{ mA}$$

$$\therefore \quad R_S = \frac{V_{MAX} - V_D}{I_{SMAX}} = \frac{22 \text{ V} - 9.1 \text{ V}}{168 \text{ mA}} = 77 \ \Omega$$

However, the power rating for this resistor is given by

$$P_{RS} = (I_{SMAX})^2 \times R_S = (168 \text{ mA})^2 \times 77 \ \Omega = 2.2 \text{ W}$$

Not only does this produce a more expensive solution, as the cost of resistors depends on the power rating, but also at the maximum supply voltage the diode is operating at its limits – this is to be avoided if at all possible.

Case 1 produces the ‘best’ engineering solution.

1.7 BS 1852 resistance code

British Standard BS 1852:1975 introduced a new code to indicate the value of many fixed and variable resistors. It uses few characters and should avoid confusion over the position of decimal points. According to this code,

0.47 Ω would be marked R47
1 Ω would be marked 1R0
4.7 Ω would be marked 4R7
47 Ω would be marked 47R
100 Ω would be marked 100R
1 kΩ would be marked 1K0
10 kΩ would be marked 10K
10 MΩ would be marked 10M

and so on.

After this code, a letter may be added to indicate the tolerance. Resistance values on diagrams in this book have been given using this standard, though not for tolerances.

Problems

1.1 The 'valency' of a semiconductor material is ....

1.2 An n-type semiconductor has an excess of ....

1.3 Electrons flow from the .... to the .... in a diode.

1.4 In a diode, conventional current flows

a) from negative to positive.
b) from positive to negative.
c) in the opposite direction to electron flow.
d) (b) and (c) above.
e) none of the above.

1.5 A diode

a) is a perfect one-way valve and stops the flow of all electrons in the reverse direction.
b) does not stop all electrons leaking in the reverse direction.
c) a device through which no current flows when the anode and cathode voltages are equal.
d) none of the above.
e) (a) and (c) above.
f) (b) and (c) above.

1.6 If a silicon diode is carrying a 10 mA current, the forward volt drop is less than

a) 0.8 V.
b) 0.2 V.
c) 25 V.
d) 50 μA.
e) (a) and (c).
f) none of the above.

1.7 The reverse breakdown voltage of a silicon diode is less than

a) 250 V.
b) 1 V.
c) 1 μA.
d) 75 V.
e) none of the above.

1.8 Wasted electrical energy in a diode is dissipated as

a) vibration.
b) heat.
c) excessive current flow.
d) resistance.
e) none of the above.

1.9 The minority carriers in p-type semiconductor are .... and are caused by ....

1.10 Assuming that the semiconductor material has zero resistance, find the current in each of the circuits of fig. 1.14.

1.11 Assuming that the zener diodes in fig. 1.15 are ideal ($I_H = 0$), calculate the maximum possible output current I_L before the output voltage falls below the stabilised value.

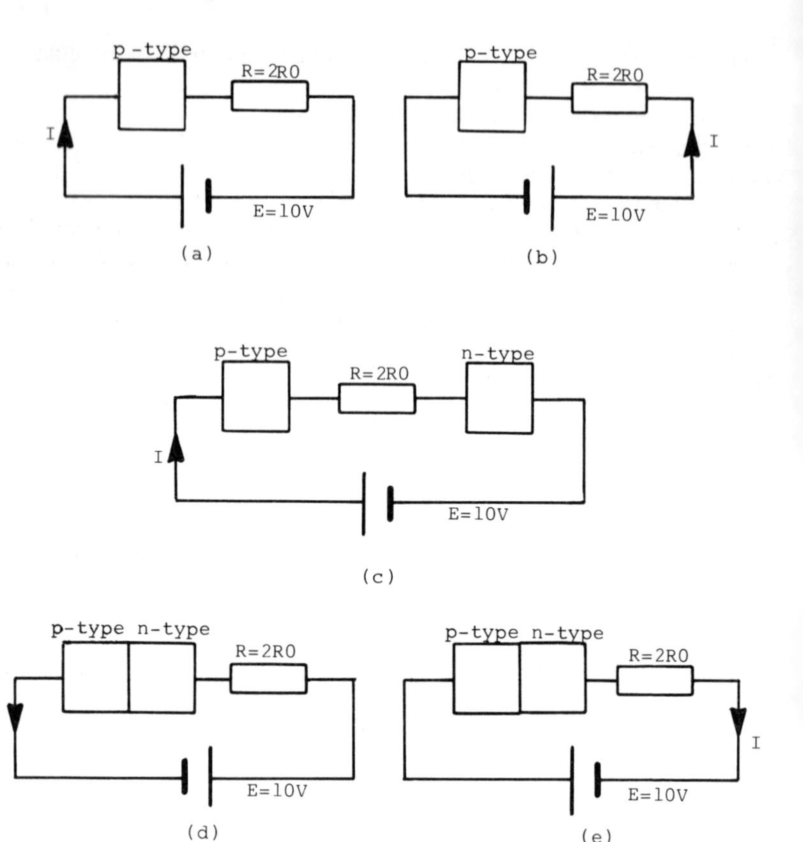

Fig. 1.14 Circuits for problem 1.10

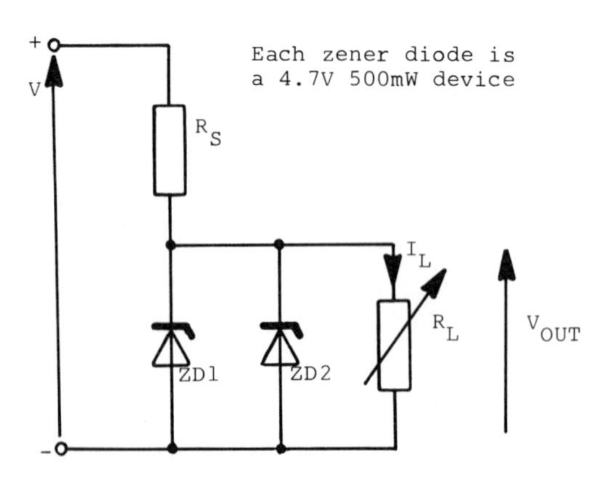

Fig. 1.15 Circuit for problem 1.11

1.12 A zener diode differs from an ordinary silicon diode because
a) it is more heavily doped.
b) it has no minority carriers.
c) it cannot conduct in the forward direction.
d) it is not affected by temperature.
e) the reverse breakdown voltage is higher.

1.13 A 24 V 500 mW zener diode is to be used to provide a stabilised supply to a variable load. The input voltage is constant at 30 V. Calculate
a) a suitable resistance value and power rating for a series limiting resistor (hint – you must decide on maximum and minimum values for the load);
b) the diode current when $R_L = 2\ k\Omega$.

1.14 A 10 V 1.3 W zener diode is to be used in a voltage stabiliser. The input voltage may vary between 22 V.and 26 V. The load is constant at 25 mA, and the minimum diode current is 4 mA. Calculate
a) the value and rating of the series resistor,
b) the power dissipated by the diode when the input voltage is 22 V.

1.15 Calculate the maximum current each of the following diodes can carry:
a) a 6.8 V 500 mW zener diode,
b) a 12 V 500 mW zener diode.

1.16 For each of the following series of zener diodes, determine the power rating:
a) BZX85,
b) BZY88.

2 Bipolar junction transistors

2.1 Introduction

What is a 'bipolar transistor', usually referred to simply as a 'transistor'? The term 'bipolar' indicates that two current carriers – holes and electrons – are involved in the operation of the device. Field-effect transistors (see chapter 4) are 'unipolar' devices. The name 'transistor' is a shortened version of 'transfer resistor', the name given by the original inventors – Schockley, Bardeen, and Brittain – since the device seemed to transfer current from a low-resistance input circuit to a high-resistance output circuit.

Like the semiconductor diode, the transistor is made from a doped crystal of germanium or silicon. Most modern transistors are made from silicon. The junction transistor is a three-layer sandwich formed with n-type and p-type semiconductor, which means that two types of transistor can be made – the npn transistor and the pnp transistor. Figure 2.1 shows the physical construction and symbols for the two types.

The three electrodes are called the *emitter, collector*, and *base*. The currents flowing in the electrodes are generally referred to as I_E, I_C, and I_B, the subscript indicating which electrode we are referring to.* Note that the circle around the symbol is often omitted, but strictly this should be done only for transistors within integrated circuits.

2.2 Transistor action

For normal transistor operation, the junctions of the transistor must be correctly biased. The biasing requirements are

a) the collector–base (or just the 'collector') junction is *reverse*-biased. This means that the junction resistance is very high.

* One convention adopted uses subscripts to indicate which of the various currents or voltages is being referred to. In the case of currents there is little problem in interpreting what is meant, but in the case of voltages a little explanation is useful. Where a single subscript is used, it is taken to mean a voltage measured between the terminal 'named' by the subscript and the ground or earth rail; e.g. V_B is the voltage measured between the base of a transistor and the ground and V_{BE} is the voltage measured between the base and emitter of a transistor. V_{CC} is usually used for the supply voltage. To make matters more difficult, upper-case (capital) subscripts are generally used to indicate d.c. or r.m.s. quantities and lower-case for instantaneous a.c. values. All a.c. values are assumed to be r.m.s. *unless* otherwise stated. To avoid confusion, we will use capital letters and subscripts for currents and voltages.

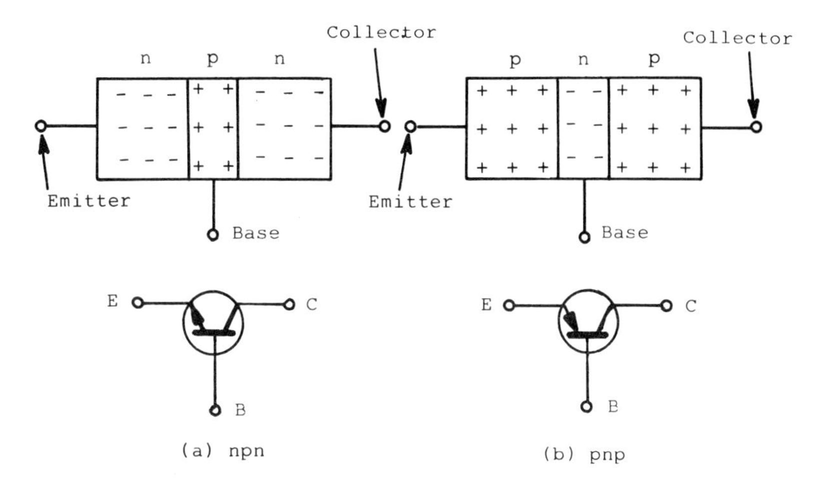

Fig. 2.1 Transistor structure and symbols

b) the emitter–base (or just the 'emitter') junction is *forward*-biased. In this case the junction resistance is low.

Figure 2.2(a) shows the biasing arrangements and conventional-current directions in an npn transistor.

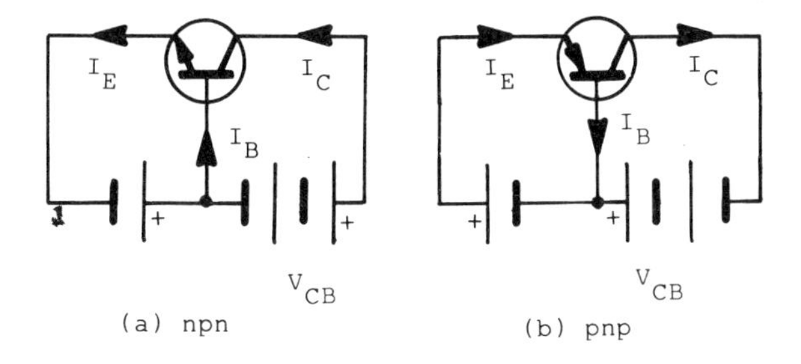

Fig. 2.2 Common-base arrangement: (a) npn, (b) pnp

The device in fig. 2.2(a) is connected in common-base mode (or common-base configuration). This means that the input to the circuit will be connected between the emitter and the base, while the output from the circuit will be taken from the collector and the base, i.e. the base is 'common' to both input and output.

Operation of the npn device can be explained as follows. The emitter junction is forward-biased, so a large number of electrons will be 'injected' into the base region, where they become minority carriers. Once in the base region, the minority carriers diffuse throughout the region and a small number recombine with holes. The majority of electrons which do not recombine are swept across the 'high-resistance' reverse-biased collector junction to form the collector current. What has happened is that we have started a large current flowing in the 'low-resistance' emitter circuit with a small voltage and have made this current flow in the 'high-resistance' collector circuit. Those electrons which recombine in the base form the small base current.

The reasons why the majority of electrons injected into the base cross the collector junction are

i) the large positive potential of the collector-junction biasing battery, which attracts electrons;
ii) the emitter region is very heavily doped, i.e. it has many free electrons, and the base region is lightly doped, i.e. it does not have many available holes for the electrons injected from the emitter to 'fall' into and recombine;
iii) the base region is made very thin, and so those electrons which do not recombine are quickly swept across the collector junction.

From a study of fig. 2.1 it is not obvious why one end of the sandwich is called the emitter and the other the collector. In fact, these two terminals can be swapped and the device will still function as a transistor; however, it will not be as efficient – the gain (see below) will be lower. The reasons for the difference are practical, not theoretical; i.e. the emitter not the collector is heavily doped, and the collector volume is much larger than that of the emitter since it has to dissipate much more power (because the current is the same – or nearly – in both, but the collector circuit has a much higher resistance and hence the 'I^2R' value is higher).

Applying Kirchoff's first law, we can see that

$$I_E = I_C + I_B$$

This formula is true for all transistors and circuit arrangements. As I_B is very small (in practice, over 99% of all electrons injected into the base cross into the collector), for most work with transistors the approximation $I_C = I_E$ is adequate – it is only in the case of power transistors that this may not be acceptable.

Any change in I_E causes a change in I_C. The ratio of the change in I_C to the change in I_E causing it is the current amplification, or gain, and is called the 'static value of the short-circuit forward current transfer ratio', h_{FB}. The 'current transfer ratio' is simply the ratio of the change in output current to the change in input current, in this case $\Delta I_C/\Delta I_E$, where Δ indicates a small change. It is referred to as 'static' because the values used to obtain the changes are taken with fixed d.c. conditions, rather than with changing ones. In practice, it is not necessary to use the change between two sets of static values, since any static (or d.c.) value for I_E and its corresponding value of I_C give approximately the same answer.

The quantity h_{FB} is one of the 'h' or hybrid parameters which are used to explain the operation of the transistor mathematically. The first subscript gives the parameter concerned e.g. 'F' for current gain; the second tells us how the transistor is connected – the 'B' above indicates that the transistor is connected in common-base mode. Upper-case (capital) subscripts are used to indicate that the measurements were taken under d.c. conditions. Lower-case subscripts are used to indicate that they apply to a.c. or 'changing' conditions e.g. h_{fb}. These are discussed in more detail in section 2.6.

Since h_{FB} varies with the load in the collector circuit, it is usually measured with the collector–base voltage constant.

Thus $h_{FB} = \dfrac{\Delta I_C}{\Delta I_E}$

and $\Delta I_C = h_{FB} \Delta I_E$

$\therefore \quad h_{FB} < 1$ since I_C is always less than I_E.

The operation of a pnp transistor is similar to that of the npn device. In this case all batteries or power supplies would be reversed and holes would be injected from the emitter into the base. The basic circuit arrangement is shown in fig. 2.2(b). The majority of transistors today are npn devices; this is because they are easier to manufacture and they are able to operate at higher frequencies than pnp types because the 'electron' moves faster than the 'hole'.

There are two more alternative and useful ways a transistor can be connected into a circuit. The identification is again derived from the lead which is common to both input and output circuits. The first, the most convenient and widely used, is the common-emitter configuration (see section 2.3). The other is the common-collector or emitter-follower configuration (section 2.4).

2.2.1 Transistor characteristics

A convenient way of presenting much of the data relating to transistors is to display the information graphically. These graphs are referred to as 'characteristics'. Three of the most useful are the input, output, and transfer characteristics.

2.2.2 Common-base input characteristic

Figure 2.3(a) shows the common-base input characteristic for a silicon transistor; this is simply a graph of the input current against the input voltage V_{BE}. Curves may be given for various values of V_{CB}. If you observe this characteristic carefully, you should see that it is identical to the diode characteristic, for the input circuit is simply a diode. Understanding the importance of this fact can be invaluable when fault-finding in electronic circuits.

2.2.3 Common-base output characteristics

Figure 2.4 shows typical common-base output characteristics for various values of I_E.

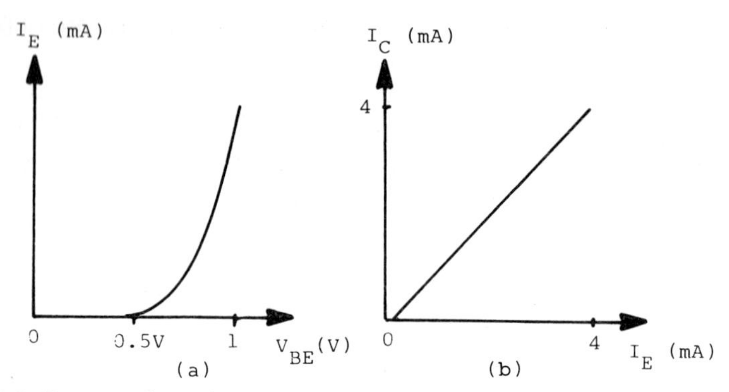

Fig. 2.3 Common-base characteristics for a silicon device: (a) input, (b) transfer

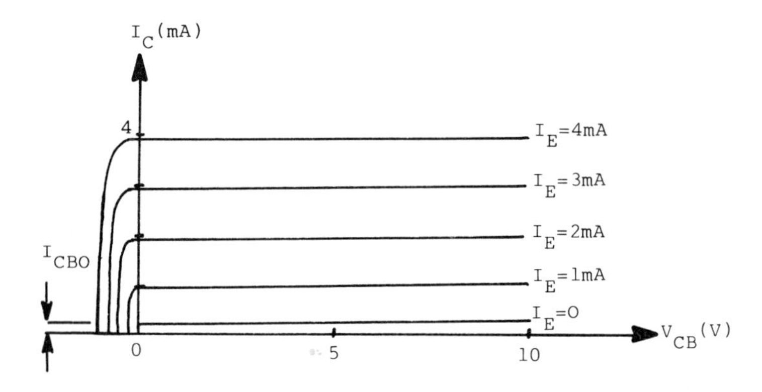

Fig. 2.4 Common-base output characteristics for a silicon device

From the characteristics of fig. 2.4, we can deduce some useful facts:

a) Each characteristic is almost horizontal, indicating a high output resistance.

b) I_C is virtually independent of V_{CB}, thus the transistor is a constant-current device. Note that I_C flows even when V_{CB} is zero; this is because any emitter current flowing in the base sets up a small reverse bias in the collector junction.

c) I_C is almost completely dependent on I_E – the device is current-operated.

d) For negative values of V_{CB}, the collector junction is forward-biased, thus a large current would flow.

e) When I_E is zero, i.e. the emitter is open circuit, a small collector current flows. This is a 'leakage' current caused by thermally generated electron–hole pairs in the collector junction. This is given the symbol

I_{CBO} – the current between the collector and base with the current in the third electrode (the emitter in this case) zero.

2.2.4 Common-base transfer characteristic

Figure 2.3(b) shows the transfer characteristic, which is simply the relationship between the input and output currents.

The various characteristics can be determined using simple circuit arrangements. This is covered in the appendix.

2.3 Common-emitter configuration

Figure 2.5 shows the basic common-emitter connection and conventional current directions. The input is now the base current. Small changes in I_B will result in a much larger change in collector current I_C. The ratio of the change in I_C to the change in I_B causing it is the current amplification, or gain, and is defined as the 'static value of the short-circuit forward current transfer ratio', h_{FE}. The first subscript indicates the parameter, current gain; the second, 'E', that the transistor is connected in common-emitter mode.

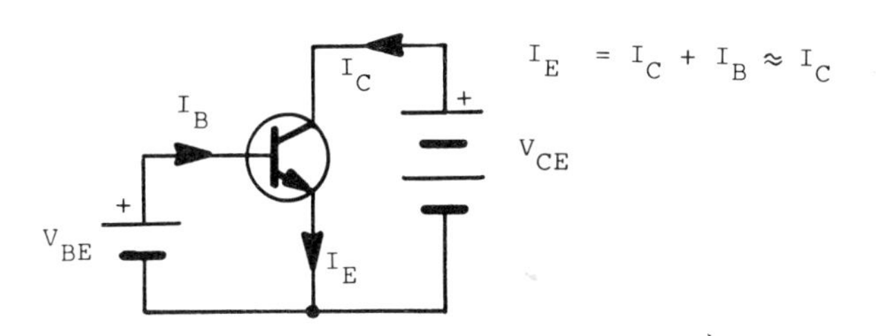

Fig. 2.5 Common-emitter arrangement

Since h_{FE} varies with I_C, it is usually quoted at the particular value of I_C at which it was measured. As with h_{FB}, an approximate value for h_{FE} can be found using a static value of the input current I_B and the corresponding value of the output current I_C. In practice, it is more convenient to use the ratio of small changes in the two currents.

$$\therefore \quad h_{FE} = \frac{\Delta I_C}{\Delta I_B}$$

$$\text{Now} \quad \Delta I_E = \Delta I_C + \Delta I_B$$

$$\therefore \quad \Delta I_B = \Delta I_E - \Delta I_C$$

$$\therefore \quad \Delta h_{FE} = \frac{\Delta I_C}{\Delta I_E - \Delta I_C}$$

Dividing each term by ΔI_E gives

$$h_{FE} = \frac{\Delta I_C/\Delta I_E}{\Delta I_E/\Delta I_E - \Delta I_C/\Delta I_E}$$

but $\Delta I_C/\Delta I_E = h_{FB}$ as has already been shown

$$\therefore \quad h_{FE} = \frac{h_{FB}}{1 - h_{FB}}$$

which gives the relationship between the common-emitter and common-base current gains.

The value of h_{FE} can be from about 10 up to 1000, depending on the type of transistor, and can be found from the manufacturer's data sheets.

The general mathematical representation is

$$I_C = h_{FE} I_B$$

2.3.1 Common-emitter input characteristic

Figure 2.6(a) shows a typical input characteristic for the common-emitter configuration. Again this can be seen to be the 'diode' characteristic.

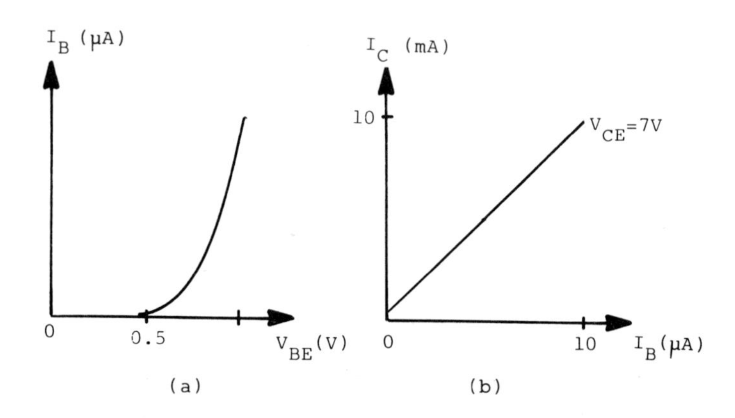

Fig. 2.6 Common-emitter characteristics: (a) input, (b) transfer

2.3.2 Common-emitter output characteristics

Figure 2.7 shows typical common-emitter output characteristics for various values of I_B. Again we can deduce a number of facts:

a) The characteristics are not quite as level as for the common-base configuration, indicating that the output resistance is not as high.
b) Collector current is virtually independent of the collector voltage and dependent on base current.
c) When V_{CE} is zero, the collector current is now zero.

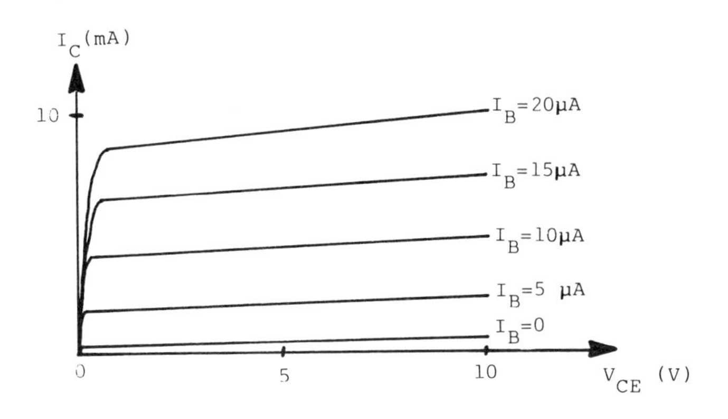

Fig. 2.7 Common-emitter output characteristics

2.3.3 Common-emitter transfer characteristic

Figure 2.6(b) shows a typical common-emitter transfer characteristic.

2.4 Common-collector (emitter-follower) configuration

The third basic transistor arrangement – the common-collector or emitter-follower configuration – is shown in fig. 2.8. This circuit arrangement has a rather restricted use, its main application being as a 'buffer' amplifier, which is an amplifier with a gain of one, used to isolate one circuit from another to prevent one circuit affecting the other. This will not be discussed further here.

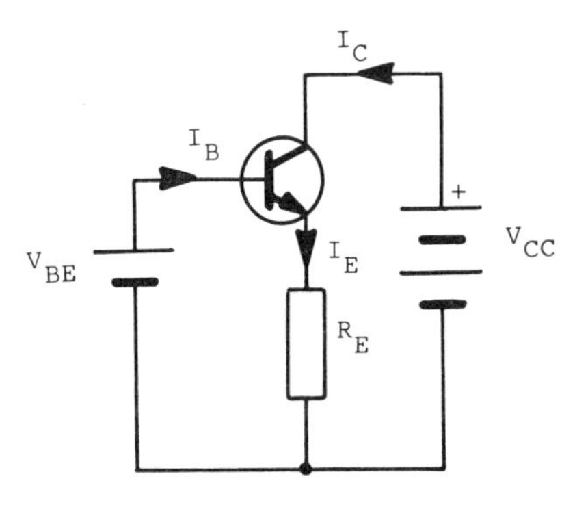

Fig. 2.8 Common-collector (emitter-follower) arrangement

2.5 Comparison of the three modes of operation

Table 2.1 compares the most important features of the three modes of transistor operation.

Table 2.1 The three modes of transistor operation

Feature	Configuration		
	Common-base	Common-emitter	Common-collector
Power gain	Yes	Yes – highest	Yes
Voltage gain	Yes	Yes	No
Current gain	No	Yes	Yes
Input impedance	Lowest, ≈ 50 Ω	Medium, ≈ 1 kΩ	Highest, ≈ 300 kΩ
Output impedance	Highest, ≈ 1 MΩ	Medium, ≈ 50 kΩ	Lowest, ≈ 300 Ω

2.6 Transistor parameters and characteristics

The basic transistor, most often in common-emitter mode, when connected with other suitably arranged components can be made into an amplifier, as discussed in chapter 3. The operation of the amplifier depends on such factors as its gain, its input and output impedances, and where on the characteristics the transistor is to be operated. In practice, a transistor is usually operated over a small linear part of its characteristics.

To define completely the operation of a transistor, only four parameters are needed. These parameters enable simple equivalent circuits to be used to represent the transistor, allowing mathematical solutions to be obtained for circuit design. Various sets, such as the 'T-parameters', have been used but, due to the difficulties in measuring the values, the semiconductor industry has adopted two sets of parameters – the 'hybrid' or 'h' parameters and the 'hybrid-pi'. Except for analysis at very high frequencies, the simple 'h' parameters are adequate for most purposes and are easily obtained from the manufacturer's characteristics. We will only consider these here. The term 'hybrid' is used because of the use of 'mixed' units, e.g. input resistance and output admittance. (Admittance – the reciprocal of impedance – is used to make the mathematical equations for parallel circuits simpler. For our applications, the impedance can be considered to be wholly resistive. The unit for admittance is the siemen, S.)

The hybrid parameters can be defined as

h_F = the static short-circuit forward current transfer ratio. This has no units – it is just a number.

h_I = the static short-circuit input resistance – units ohms.

These two parameters are obtained with the output short-circuited, $V_2 = 0$ in fig. 2.9(a); hence 'short-circuit' in the definition.

h_R = the static value of the reverse voltage feedback ratio and is a pure number. This indicates the effect that the output voltage has on the input voltage. For all practical purposes this can be ignored.

h_O = the static output admittance – units siemens. For most purposes this can also be ignored.

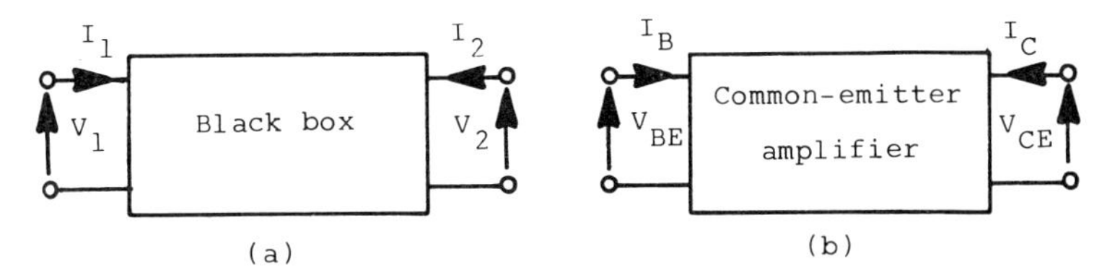

Fig. 2.9 'Black box' concept: (a) general, (b) common-emitter

These last two parameters are obtained with the input open circuit, $I_1 = 0$ in fig. 2.9(a).

The static values for the parameters can be obtained from the characteristics.

Example 2.1 Figure 2.10 shows the common-base input characteristic for a BC109 transistor. Determine a value for h_{IB}.

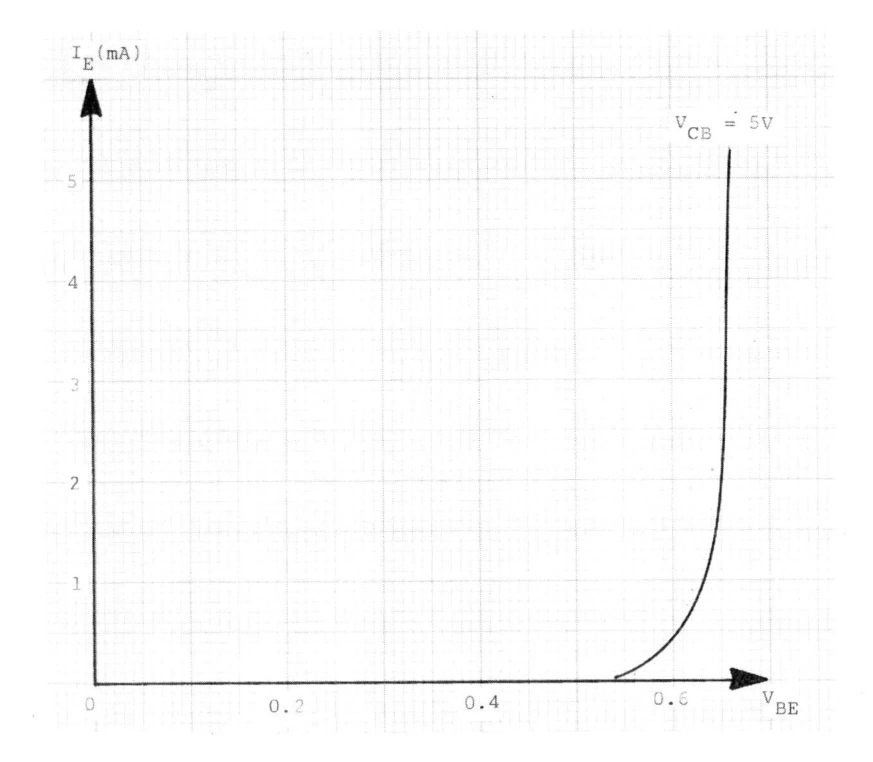

Fig. 2.10 Characteristics for example 2.1

1. First draw a tangent to the linear part of the characteristic.
2. Determine $\Delta V_{BE} = 0.66\text{ V} - 0.645\text{ V} = 0.15\text{ V}$.
3. Determine $\Delta I_E = 5.3\text{ mA} - 0 = 5.3\text{ mA}$.
4. Calculate $h_{IB} = \dfrac{\Delta V_{BE}}{\Delta I_E} = \dfrac{15\text{ mV}}{5.3\text{ mA}} = 2.83\ \Omega$

Example 2.2 Figure 2.11 shows the common-base output characteristic for a BC109 transistor. Determine h_{FB} for $V_{CB} = 6$ V and h_{OB} for $I_E = 3$ mA.

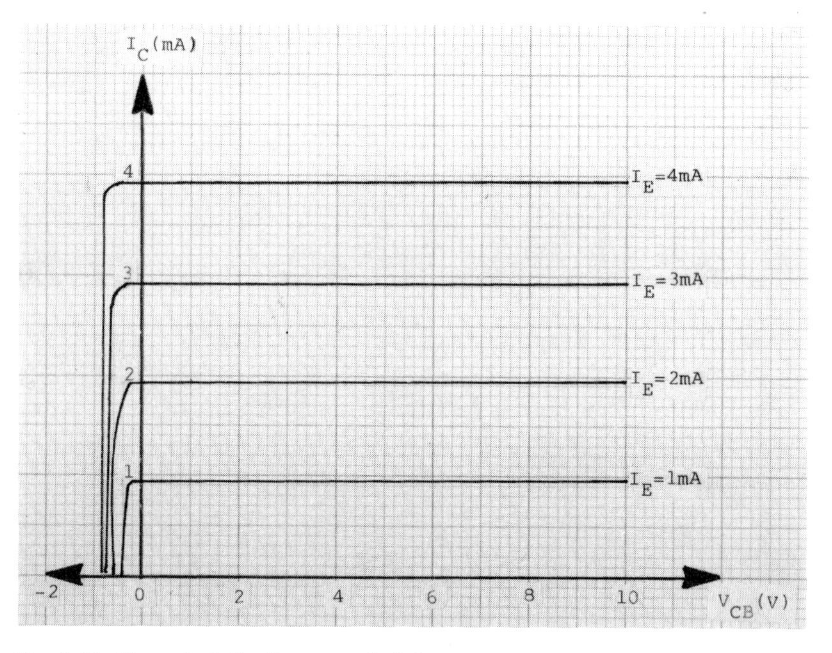

Fig. 2.11 Characteristics for example 2.2

For h_{FB}:

1. Draw a vertical line at $V_{CB} = 6$ V.
2. Determine $\Delta I_E = 4\text{ mA} - 1\text{ mA} = 3\text{ mA}$.
3. Determine $\Delta I_C = 3.9\text{ mA} - 0.95\text{ mA} = 2.95\text{ mA}$.
4. Calculate $h_{FB} = \dfrac{\Delta I_C}{\Delta I_E} = \dfrac{2.95\text{ mA}}{3\text{ mA}} = 0.98$.

For h_{OB}:

1. Draw as large a triangle as possible over the linear part of the $I_E = 3$ mA characteristic.
2. Determine $\Delta V_{CB} = 10\text{ V} - 0 = 10\text{ V}$.
3. Determine $\Delta I_C = 2.95\text{ mA} - 2.9\text{ mA} = 0.05\text{ mA}$.
4. Calculate $h_{OB} = \dfrac{\Delta I_C}{\Delta V_{CB}} = \dfrac{0.05\text{ mA}}{10\text{ V}} = 5\ \mu\text{S}$.

Example 2.3 Figure 2.12 shows the input characteristic for a BC109 transistor in common-emitter mode. Determine a value for h_{IE}.

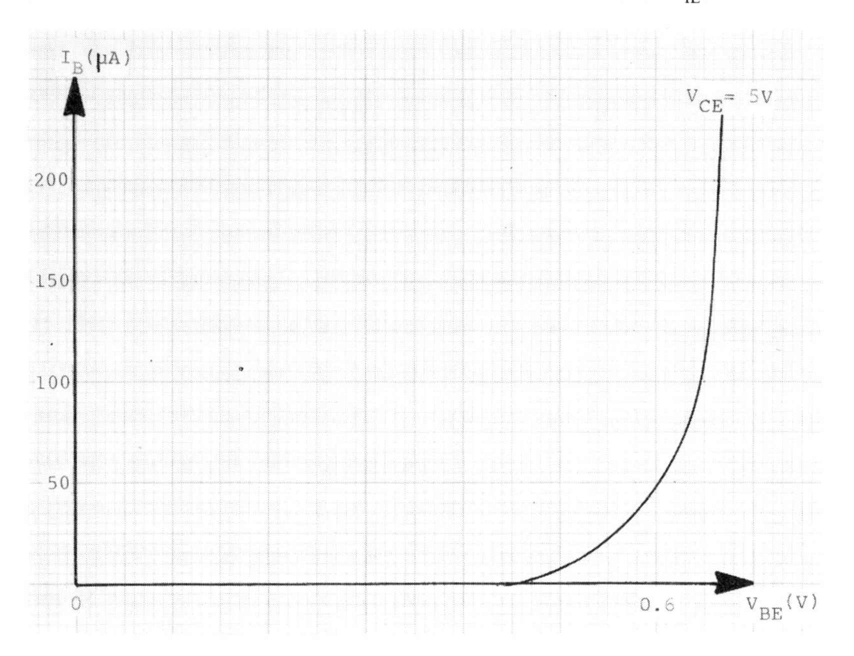

Fig. 2.12 Characteristics for example 2.3

1. Draw a tangent to the linear part of the characteristic.
2. Determine $\Delta V_{BE} = 0.7\text{ V} - 0.4\text{ V} = 0.3\text{ V}$.
3. Determine $\Delta I_B = 235\ \mu\text{A} - 0 = 235\ \mu\text{A}$.
4. Calculate $h_{IE} = \dfrac{\Delta V_{BE}}{\Delta I_B} = \dfrac{0.3\text{ V}}{235\ \mu\text{A}} = 1.28\text{ k}\Omega$.

Example 2.4 Figure 2.13 shows the output characteristics for a BC109 transistor in common-emitter mode. Determine h_{FE} for $V_{CE} = 6$ V and h_{OE} for $I_B = 150$ μA.

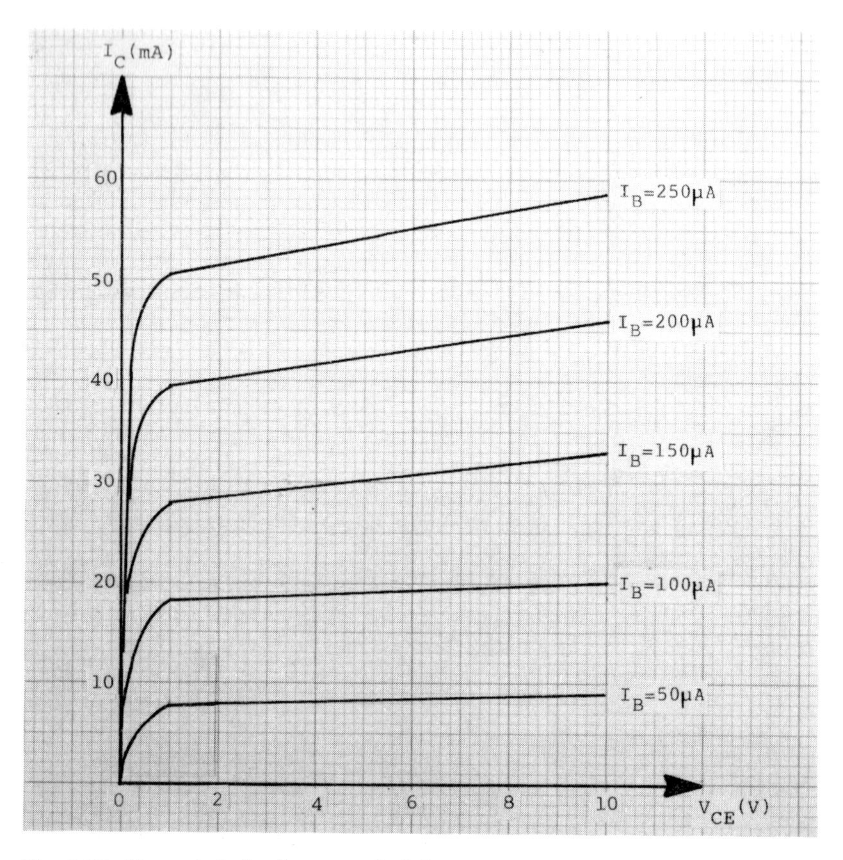

Fig. 2.13 Characteristics for example 2.4

For h_{FE}:

1. Draw a vertical line at $V_{CE} = 6$ V.
2. Determine $\Delta I_B = 250$ μA $- 50$ μA $= 200$ μA.
3. Determine $\Delta I_C = 55$ mA $- 8$ mA $= 47$ mA.
4. Calculate $h_{FE} = \dfrac{\Delta I_C}{\Delta I_B} = \dfrac{47 \text{ mA}}{200\ \mu\text{A}} = 235.$

For h_{OE}:

1. Draw as large a triangle as possible over the linear part of the $I_B = 150$ μA characteristic.

2. Determine $\Delta V_{CE} = 10\text{ V} - 1\text{ V} = 9\text{ V}$.

3. Determine $\Delta I_C = 33\text{ mA} - 27.5\text{ mA} = 5.5\text{ mA}$.

4. Calculate $h_{OE} = \dfrac{\Delta I_C}{\Delta V_{CE}} = \dfrac{5.5\text{ mA}}{9\text{ V}} = 611\ \mu\text{S}$.

The static parameters are of limited practical use once the d.c. operating conditions have been set up – a process known as 'biasing', which we will deal with in chapter 3 – we are then only interested in the amplifier's effects on any a.c. signal we may apply to the amplifier. To prevent distortion being introduced, the a.c. signals are restricted to small changes about the operating point. The a.c. parameters are referred to as the 'small-signal' parameters, and this is indicated by using small subscripts, e.g. h_{fe} is the small-signal short-circuit forward current transfer ratio in the common-emitter mode.

The small-signal parameters can be found by taking measurements external to the transistor itself. This is done by considering the transistor to be a four-terminal 'black-box' network as shown in fig. 2.9(b). By measuring the two currents and two voltages shown, all four parameters can be found – what is inside the 'black box' can be a transistor in any mode.

Assuming a common-emitter amplifier, the two most important 'small-signal' 'h' parameters are defined below, using the circuit of fig. 2.9(b).

$$h_{fe} = \frac{\Delta I_C}{\Delta I_B} \quad \text{with } V_{CE} = 0$$

$$h_{ie} = \frac{\Delta V_{BE}}{\Delta I_B} \quad \text{with } V_{CE} = 0$$

However, this information can be obtained from the manufacturer's data sheets. Figures 2.14(a) to (d) show the data sheets for the BC107–BC109 family of devices.

Problems

2.1 For the characteristics shown in fig. 2.15, calculate
a) h_{FE} when $V_{CE} = 6$ V,
b) R_{OUT} for $I_B = 100\ \mu$A (note: $R_{OUT} = 1/h_{oe}$),
c) h_{ie} for $V_{CE} = 4.5$ V.

2.2 For the characteristics shown in figs 2.10 and 2.11, calculate
a) h_{FB} when $V_{CB} = 5$ V,
b) R_{OUT} for $I_E = 4$ mA (note: $R_{OUT} = 1/h_{ob}$),
c) h_{ib}.

2.3 The base current of an npn transistor consists mainly of
a) holes.
b) electrons.
c) equal numbers of holes and electrons.

A.F. SILICON PLANAR EPITAXIAL TRANSISTORS

N-P-N transistors in TO-18 metal envelopes with the collector connected to the case.

The **BC107** is primarily intended for use in driver stages of audio amplifiers and in signal processing circuits of television receivers.

The **BC108** is suitable for multitude of low-voltage applications e.g. driver stages or audio preamplifiers and in signal processing circuits of television receivers.

The **BC109** is primarily intended for low-noise input stages in tape recorders, hi-fi amplifiers and other audio-frequency equipment.

QUICK REFERENCE DATA

			BC107	BC108	BC109	
Collector-emitter voltage ($V_{BE} = 0$)	V_{CES}	max.	50	30	30	V
Collector-emitter voltage (open base)	V_{CEO}	max.	45	20	20	V
Collector current (peak value)	I_{CM}	max.	200	200	200	mA
Total power dissipation up to $T_{amb} = 25\ ^{o}C$	P_{tot}	max.	300	300	300	mW
Junction temperature	T_j	max.	175	175	175	^{o}C
Small-signal current gain at $T_j = 25\ ^{o}C$; $I_C = 2$ mA; $V_{CE} = 5$ V; f = 1 kHz	h_{fe}	$>$	125	125	240	
		$<$	500	900	900	
Transition frequency at f = 35 MHz; $I_C = 10$ mA; $V_{CE} = 5$ V	f_T	typ.	300	300	300	MHz
Noise figure at $R_S = 2$ kΩ; $I_C = 200\ \mu A$; $V_{CE} = 5$ V; f = 30 Hz to 15 kHz	F	typ.	–	–	1,4	dB
		$<$	–	–	4,0	dB
f = 1 kHz; B = 200 Hz	F	typ.	2	2	1,2	dB

MECHANICAL DATA

Dimensions in mm

Fig. 1 TO-18.

Collector connected to case

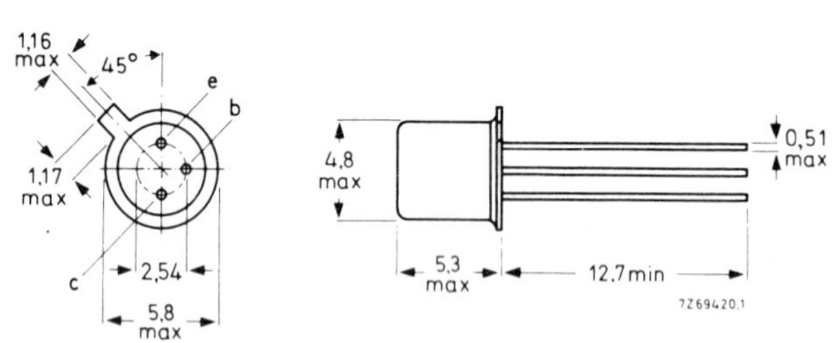

Accessories: 56246 (distance disc).

Fig. 2.14(a) Data sheets for BC107–BC109 devices

RATINGS Limiting values in accordance with the Absolute Maximum System (IEC 134)

Voltages			BC107	BC108	BC109	
Collector-base voltage (open emitter)	V_{CBO}	max.	50	30	30	V
Collector-emitter voltage (V_{BE} = 0)	V_{CES}	max.	50	30	30	V
Collector-emitter voltage (open base)	V_{CEO}	max.	45	20	20	V
Emitter-base voltage (open collector)	V_{EBO}	max.	6	5	5	V

Currents				
Collector current (d.c.)	I_C	max.	100	mA
Collector current (peak value)	I_{CM}	max.	200	mA
Emitter current (peak value)	$-I_{EM}$	max.	200	mA
Base current (peak value)	I_{BM}	max.	200	mA
Power dissipation				
Total power dissipation up to $T_{amb} = 25\,^{o}C$	P_{tot}	max.	300	mW
Temperatures				
Storage temperature	T_{stg}	−65 to	+175	^{o}C
Junction temperature	T_j	max.	175	^{o}C

THERMAL RESISTANCE

From junction to ambient in free air	$R_{th\ j\text{-}a}$	=	0.5	$^{o}C/mW$
From junction to case	$R_{th\ j\text{-}c}$	=	0.2	$^{o}C/mW$

CHARACTERISTICS $T_j = 25\,^{o}C$ unless otherwise specified

Collector cut-off current				
$I_E = 0$; $V_{CB} = 20$ V; $T_j = 150\,^{o}C$	I_{CBO}	<	15	μA
Base-emitter voltage[1]				
$I_C = 2$ mA; $V_{CE} = 5$ V	V_{BE}	typ.	620	mV
		550 to	700	mV
$I_C = 10$ mA; $V_{CE} = 5$ V	V_{BE}	<	770	mV

[1]) V_{BE} decreases by about 2 mV/^{o}C with increasing temperature.

Fig. 2.14(b) Data sheets for BC107–BC109 devices (*continued*)

CHARACTERISTICS (continued) T_j = 25 °C unless otherwise specified

Saturation voltages [1])

Conditions	Symbol		Value	Unit
I_C = 10 mA; I_B = 0.5 mA	V_{CEsat}	typ.	90	mV
		<	250	mV
	V_{BEsat}	typ.	700	mV
I_C = 100 mA; I_B = 5 mA	V_{CEsat}	typ.	200	mV
		<	600	mV
	V_{BEsat}	typ.	900	mV

Knee voltage

Conditions	Symbol		Value	Unit
I_C = 10 mA; I_B = value for which I_C = 11 mA at V_{CE} = 1 V	V_{CEK}	typ.	300	mV
		<	600	mV

Conditions	Symbol		Value	Unit
Collector capacitance at f = 1 MHz				
$I_E = I_e = 0$; V_{CB} = 10 V	C_c	typ.	2.5	pF
		<	4.5	pF
Emitter capacitance at f = 1 MHz				
$I_C = I_c = 0$; V_{EB} = 0.5 V	C_e	typ.	9	pF
Transition frequency at f = 35 MHz				
I_C = 10 mA; V_{CE} = 5 V	f_T	typ.	300	MHz

Conditions	Symbol		BC107	BC108	BC109	Unit
Small signal current gain at f = 1 kHz						
I_C = 2 mA; V_{CE} = 5 V	h_{fe}	>	125	125	240	
		<	500	900	900	
Noise figure at R_S = 2 kΩ; I_C = 200 μA; V_{CE} = 5 V						
f = 30 Hz to 15 kHz	F	typ.			1.4	dB
		<			4	dB
f = 1 kHz; B = 200 Hz	F	typ.	2	2	1.2	dB
		<	10	10	4	dB

[1]) V_{BEsat} decreases by about 1.7 mV/°C with increasing temperature.

Fig. 2.14(c) Data sheets for BC107–BC109 devices (*continued*)

CHARACTERISTICS (continued) T_j = 25 °C unless otherwise specified

			BC107A BC108A	BC107B BC108B BC109B	BC108C BC109C	
D.C. current gain						
I_C = 10 μA; V_{CE} = 5 V	h_{FE}	>		40	100	
		typ.	90	150	270	
I_C = 2 mA; V_{CE} = 5 V	h_{FE}	>	110	200	420	
		typ.	180	290	520	
		<	220	450	800	
h parameters at f = 1 kHz (common emitter)						
I_C = 2 mA; V_{CE} = 5 V						
Input impedance	h_{ie}	>	1.6	3.2	6	kΩ
		typ.	2.7	4.5	8.7	kΩ
		<	4.5	8.5	15	kΩ
Reverse voltage transfer ratio	h_{re}	typ.	1.5	2	3	10^{-4}
Small signal current gain	h_{fe}	>	125	240	450	
		typ.	220	330	600	
		<	260	500	900	
Output admittance	h_{oe}	typ.	18	30	60	$\mu\Omega^{-1}$
		<	30	60	110	$\mu\Omega^{-1}$

Fig. 2.14(d) Data sheets for BC107–BC109 devices (*continued*)

2.4 The hybrid-pi equivalent circuit is used to represent the transistor operation at .... frequencies.

2.5 Explain why a small collector current flows in a common-base-connected transistor when the collector-junction reverse bias is zero.

2.6 Explain why the majority of electrons injected into the base of a correctly biased npn transistor cross the high-resistance reverse-biased collector junction.

2.7 The transistor configuration which produces the highest power gain is ....

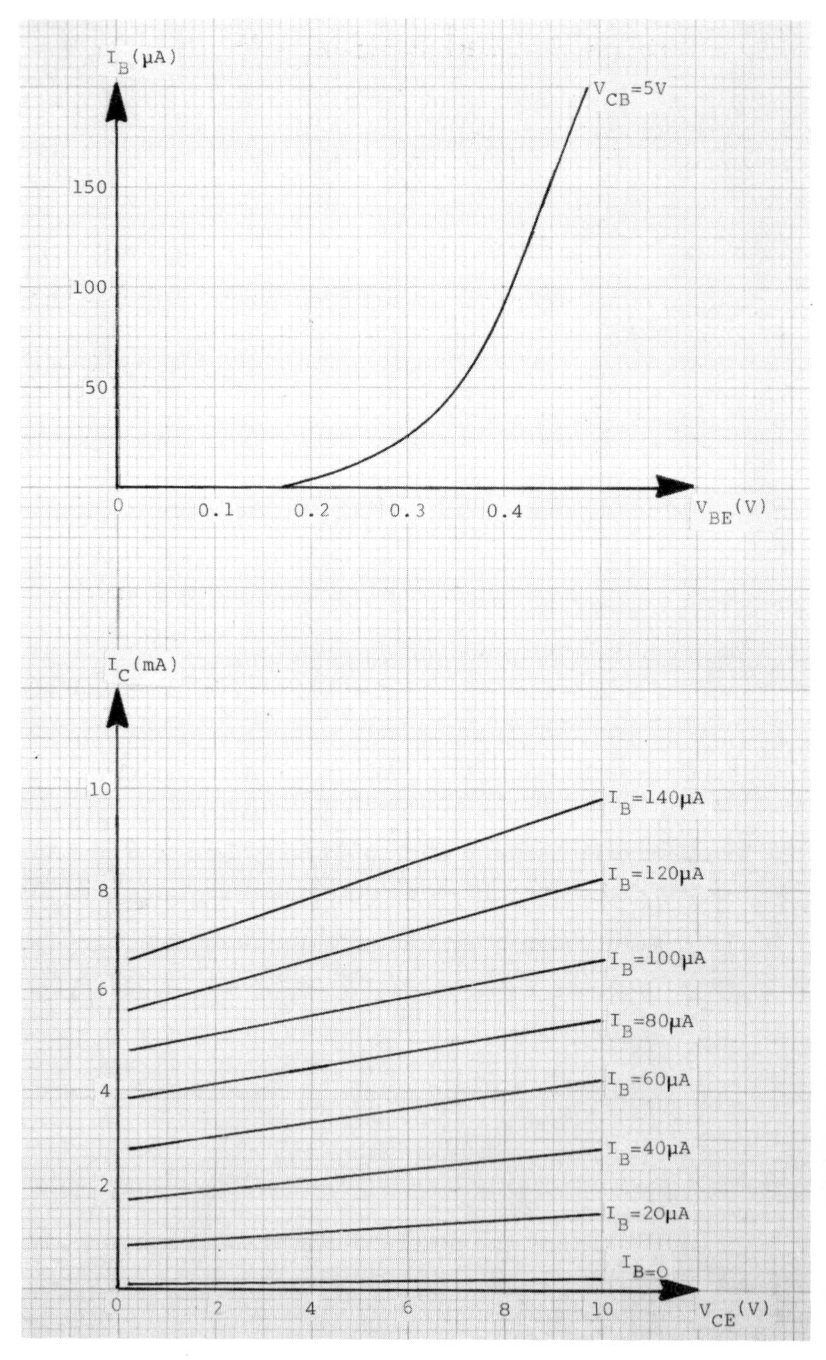

Fig. 2.15 Characteristics for problem 2.1

3 Small-signal amplifiers

3.1 Introduction

The function of an amplifier is to increase the voltage, current, or power of an input signal to the amplifier. Ideally the amplifier should do this without distorting the signal in any way. To do this, it is necessary to take the basic transistor of chapter 2 and set it up or 'bias' it so that we can achieve the required amplification. We will also look at how we can predict or calculate the gain.

The vast majority of all amplifiers are common-emitter type, and we will consider this type in detail. It is particularly convenient since we can obtain voltage, current, and power gains with it.

The applications for small-signal amplifiers are too numerous to list, but obviously the radio amplifier is one of the most easy to identify.

3.2 The common-emitter amplifier

Since, as explained in chapter 2, the bipolar transistor is a current-operated device, it is necessary to put a 'load resistor', R_C in fig. 3.1,* in

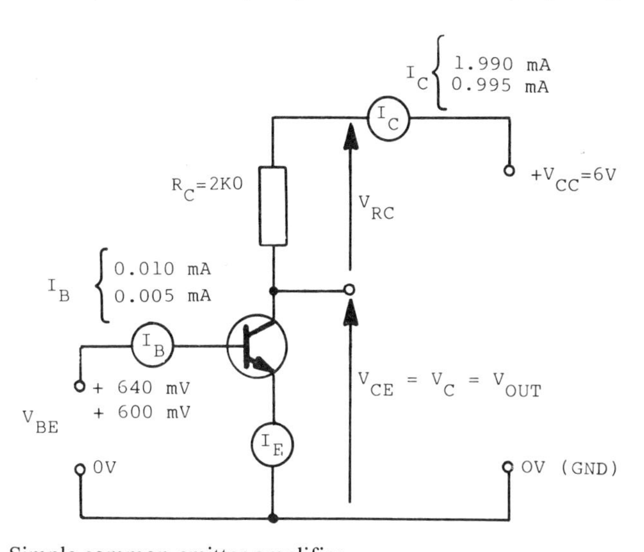

Fig. 3.1 Simple common-emitter amplifier

* It is a convenient convention that resistors connected to transistors are identified by a subscript which relates to the transistor terminal to which they are connected.

series with the collector of the transistor to change the collector current into voltage.

Consider the circuit of fig. 3.1. The collector junction is reverse-biased and therefore a very high resistance, so adding a resistor R_C of a few thousand ohms to the collector circuit will not change the collector current significantly.

The output voltage V_C (note: in this case V_C and V_{CE} are the same) can be found by applying Kirchoff's second law – i.e. it is the supply voltage minus the p.d. across the collector resistor,

$$\therefore \quad V_C = V_{CC} - V_{RC} = V_{CC} - I_C R_C \qquad 3.1$$

When the collector current I_C changes, then so must V_C.

Suppose the initial conditions for the circuit of fig. 3.1 are V_{BE} = 600 mV, causing I_B = 0.005 mA which results in I_C = 0.995 mA,

$$\text{then} \quad V_{RC} = I_C \times R_C = 0.995 \text{ mA} \times 2 \text{ k}\Omega \approx 1 \text{ mA} \times 2 \text{ k}\Omega = 2 \text{ V}$$

$$\text{thus} \quad V_C = V_{OUT} = V_{CC} - V_{RC} = 6 \text{ V} - 2 \text{ V} = 4 \text{ V}$$

If we now increase V_{BE} to 640 mV, resulting in I_B doubling to I_B = 0.01 mA and I_C changing to 1.99 mA ≈ 2 mA, then

$$V_{OUT} = V_{CC} - V_{RC} = 6 \text{ V} - 4 \text{ V} = 2 \text{ V}$$

i.e. V_{OUT} has changed by 2 V. Thus a change of 40 mV at the input has brought about a change of 2 V at the output.

The voltage amplification or gain, A_V, is the ratio of these two voltages,

$$\therefore \quad A_V = \frac{\Delta V_{OUT}}{\Delta V_{IN}} = \frac{2 \text{ V}}{40 \times 10^{-3} \text{ V}} = 50$$

(Note: this is a slight approximation.)

The initial input power is

$$V_{IN} \times I_B = 600 \text{ mV} \times 0.005 \text{ mA} = 3\ \mu\text{W}$$

The final input power is

$$V_{IN} \times I_B = 640 \text{ mV} \times 0.01 \text{ mA} = 6.4\ \mu\text{W}$$

The resulting change in input power, ΔP_{IN}, is given by

$$\Delta P_{IN} = 6.4\ \mu\text{W} - 3\ \mu\text{W} = 3.4\ \mu\text{W}$$

The initial output power is

$$V_{OUT} \times I_C = 4 \text{ V} \times 1 \text{ mA} = 4 \text{ mW}$$

The final output power is

$$V_{OUT} \times I_C = 2 \text{ V} \times 2 \text{ mA} = 8 \text{ mW}$$

The resulting change in output power, ΔP_{OUT}, is given by

$$\Delta P_{OUT} = 8 \text{ mW} - 4 \text{ mW} = 4 \text{ mW}$$

$$\text{Thus} \quad \text{power gain } A_P = \frac{\Delta P_{OUT}}{\Delta P_{IN}} = \frac{4 \text{ mW}}{3.4 \,\mu\text{W}} = 1177$$

The current gain A_I is the ratio of the change in output current to the change in input current.

$$\therefore \quad A_I = \frac{\Delta I_C}{\Delta I_B} = \frac{2 \text{ mA} - 1 \text{ mA}}{0.01 \text{ mA} - 0.005 \text{ mA}} = \frac{1 \text{ mA}}{0.005 \text{ mA}} = 200$$

Note: in this case $A_I = h_{FE}$.

3.2.1 A.C. amplification

The previous calculations used discrete or fixed values for the voltages and currents. In the case of a.c. signals which are connected from an external circuit and are probably going to be connected to an external output circuit after amplification, we must prevent any d.c. interaction between the input and output circuits. This is achieved by adding two 'd.c.-blocking' capacitors, C_{IN} and C_{OUT}, as shown in fig. 3.2.

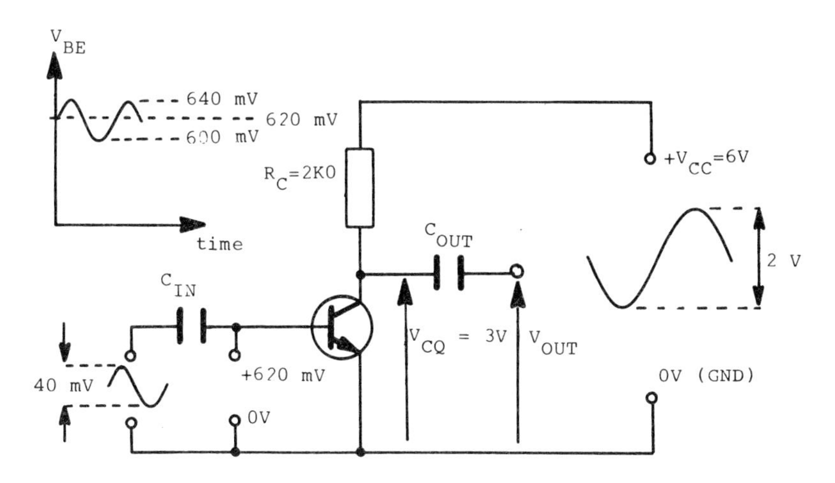

Fig. 3.2 Common-emitter a.c. amplifier

If we now apply a 40 mV peak-to-peak sine-wave input signal which is superimposed via C_{IN} on to a 620 mV base bias voltage, it will cause the base voltage to swing between 600 mV and 640 mV. Thus I_C swings between approximately 1 mA and 2 mA, and V_{OUT} swings between $\pm$ 1.0 V or 2 V peak-to-peak. The output voltage via C_{OUT} is a sine wave with no d.c. component.

The above simple example assumes that the forward current transfer ratio h_{FE} is the same for both a.c. and d.c. In practice, the a.c. value is usually a little higher than the d.c., but, provided we use small signals compared to the bias voltage – 'small-signal' amplifier – the amplifier behaves as a d.c. amplifier with small variations about the operating or quiescent point. Since the output voltage decreases when the input increases, it will be observed that the output is 180° out of phase with the input (or anti-phase).

3.2.2 Biasing the amplifier

For the amplifier of fig. 3.2 we have to arrange that there is a suitable base bias voltage so that the amplifier will operate over the linear part of its characteristics. If there is no bias, then collector current will flow only on one half cycle of the input, the positive half cycle. For small-signal amplifiers, bias is usually arranged so that with no signal applied the d.c. or quiescent collector voltage, V_{CQ}, is about half the supply voltage. (This is 'class-A' operation, which will be discussed in section 3.5 on amplifier classification.) In the case of the circuit of fig. 3.2, $V_{CQ} = 3$ V and the corresponding quiescent base current is 0.0075 mA = 7.5 μA.

Normally, however, the quiescent value of I_B is not known. One method of determining it is to use the output characteristics and a 'load line'. This is simply a straight-line graph put on the output characteristics using Kirchoff's second law.

Figure 3.3 shows the output characteristics for the transistor of fig. 3.2. Let us now put a load line on the characteristics for the circuit. The collector voltage V_C is given by equation 3.1,

i.e. $V_C = V_{CC} - I_C R_C$

Those of you who are good at maths will recognise this as an equation of the form $y = C - mx$, the equation of a straight line. Those who doubt this can plot graphs at their leisure and verify it.

To plot any straight-line graph, we need only two points. When I_C is zero, then equation 3.1 gives

$$V_C = V_{CC} - 0 \times R_C = V_{CC} - 0 = V_{CC} = 6 \text{ V}$$

When $V_C = 0$, then equation 3.1 gives

$$I_C = \frac{V_{CC}}{R_C} = \frac{6 \text{ V}}{2 \text{ k}\Omega} = 3 \text{ mA}$$

Using these two points, X = (6,0) and Y = (0,3), we can put our load line on the characteristics as shown in fig. 3.3. Note that the slope of the load line is given mathematically by the expression $-1/R_C$.

From the characteristics, we can see that when $I_B = 0$ a small collector current I_{CBO} – the thermally generated leakage current – still flows so there is a small p.d. across R_C which prevents V_C reaching V_{CC}. In practice, with modern silicon devices it is virtually impossible to measure this difference; however, the value can be found by considering the intersection of the load line and the $I_B = 0$ curve, point A on fig. 3.3.*

* For a modern silicon transistor the $I_B = 0$ curve on the characteristic is virtually coincident with the axis. That in fig. 3.3 has been exaggerated to clarify the theoretical conditions.

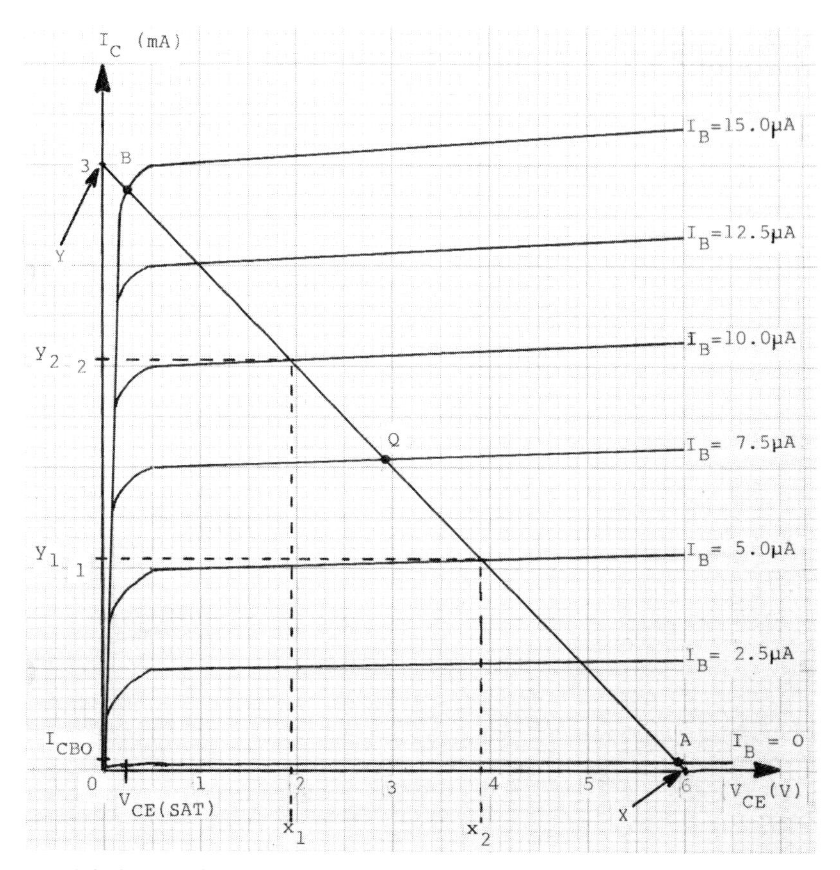

Fig. 3.3 Output characteristics for fig. 3.2

Increasing I_B increases I_C ($I_C = h_{FE} I_B$), causing V_C ($= V_{CC} - I_C R_C$) to fall. Eventually a point will be reached when $I_C \times R_C$ approaches V_{CC}; from fig. 3.3, this is when $I_B = 15\ \mu A$ and then V_C is approximately zero. V_C never falls to zero since the transistor possesses a small resistance – increasing I_B further does not result in any further increase in I_C and the transistor is said to be 'saturated' or 'bottomed'. The collector–emitter voltage in this state is called $V_{CE(SAT)}$ and can be found from the data sheet – it is typically 0.1 V to 0.2 V. This condition is shown by point B on fig. 3.3.

By varying base current between 0 and 15 μA, the d.c. quiescent or operating point, point Q, of the transistor can be set to any point along the load line from A to B. Point Q is used for class-A operation (see section 3.5 on amplifier classification), which is the case for most common-emitter amplifiers.

If we arrange for I_B to be set at 7.5 μA, the operating point will be set half-way along the load line and we have 'biased' the transistor. The d.c. base bias current will be identified by a second subscript, Q, i.e. I_{BQ}.

Even with no signal applied to the input of the transistor, there is a collector current flowing. This means that the transistor will be dissipating power P_{CQ} given by

$$P_{CQ} = I_{CQ} \times V_{CQ} \qquad 3.2$$

$$= 1 \text{ mA} \times 3 \text{ V}$$

$$= 3 \text{ mW}$$

This power is wasted in the transistor as heat.

We can now use the characteristics to calculate the gain when small signals are applied to the input.

If we use an input signal of 5 μA peak-to-peak, the base current will swing between + 10 μA and + 5 μA (i.e. 7.5 μA $\pm$ (5 μA)/2), resulting in an output current swing between y_1 and y_2, giving a ΔI_C of $y_1 - y_2 =$ 0.95 mA. (Note: the intercepts on the collector-current axis, y_1 and y_2, are obtained using the intersection of the load line and the appropriate base-current curve.) To obtain the r.m.s. value,

$$I_C = \frac{\Delta I_{C(P-P)}}{2} \times 0.707$$

$$\text{Current gain } A_I = h_{fe} = \frac{\Delta I_C}{\Delta I_B} = \frac{2.05 \text{ mA} - 1.1 \text{ mA}}{10 \ \mu\text{A} - 5 \ \mu\text{A}}$$

$$= \frac{0.95 \text{ mA}}{5 \ \mu\text{A}} = 190$$

$$\text{Voltage gain } A_V = \frac{\Delta V_C}{\Delta V_{IN}} = \frac{\Delta V_C}{\Delta I_B R_{IN}}$$

For fig. 3.3, $\Delta V_C = x_2 - x_1 = 3.9 \text{ V} - 1.95 \text{ V} = 1.95 \text{ V}$

Also, A_V can be found from

$$A_V = \frac{\Delta I_C R_C}{\Delta I_B R_{IN}} = A_I \frac{R_C}{R_{IN}}$$

R_{IN} is the input resistance of the amplifier. It may be just h_{ie}, but in general it is not. The value cannot be obtained from the characteristics and the derivation is beyond the scope of this book, so the value will be provided if required.

$$\text{Power gain } A_P = \frac{\Delta P_{OUT}}{\Delta P_{IN}} = \frac{\Delta I_C^2 R_C}{\Delta I_B^2 R_{IN}} = A_I^2 \frac{R_C}{R_{IN}}$$

3.2.3 Setting the operating point

The operating point could be set by using a separate bias battery to provide the quiescent base bias current, but this is inconvenient and expensive. The same effect can be achieved in the common-emitter amplifier by modifying the circuit of fig. 3.2 to that shown in fig. 3.4.

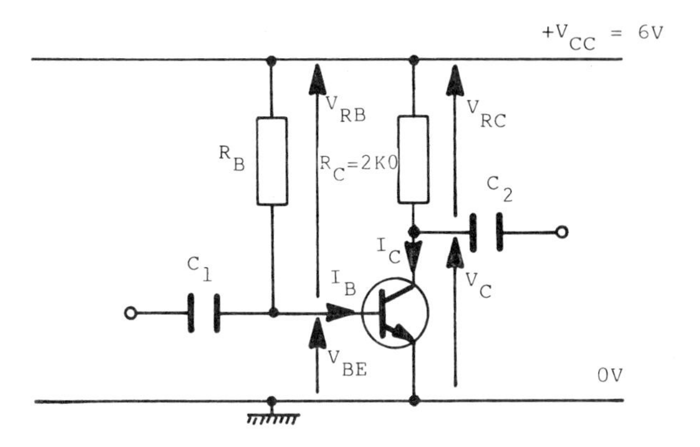

Fig. 3.4 Base-current bias

When I_{BQ} has been found as in the previous section, we can calculate a suitable value for R_B using

$$R_B = \frac{V_{CC} - V_{BE}}{I_{BQ}} = \frac{6 \text{ V} - 0.6 \text{ V}}{7.5\ \mu\text{A}} = 720 \text{ k}\Omega \qquad 3.3$$

Note: V_{BE} = 0.6 V for silicon transistors and 0.2 V for germanium.

Using Kirchoff's and Ohm's laws, it is now possible to carry out simple calculations to determine suitable component values, circuit currents, and voltages.

The simple bias arrangement in fig. 3.4 is not very satisfactory since the operating point moves with any change in I_B. Changes in I_B can be caused by variations in leakage current I_{CBO} due to temperature, transistor parameters varying, ageing, etc.

The collector current I_C consists of two components: one due to the input signal and one due to the thermally generated leakage current. If the temperature increases, so does the leakage current – causing a change in the operating point, which may cause distortion of the output signal. This distortion may be excessive. Since the transistor itself is dissipating power, the increase in I_C further increases the junction temperature which further increases I_C, making the effect cumulative.

This process can be a vicious circle leading to 'thermal runaway' which may destroy the transistor by literally melting the crystal. (This is more likely with germanium than with silicon transistors.)

Power transistors (these are usually used in the final or output stages to drive some load device, e.g. a loudspeaker) normally require heat sinks to increase the rate at which the excess heat is radiated away. The heat sinks are often 'finned' to increase the surface area and are generally painted matt black, the best colour for radiation.

An improved circuit arrangement, known as 'feedback bias stabilisation', stabilises the operating point for changes in temperature and ageing etc. and is shown in fig. 3.5.

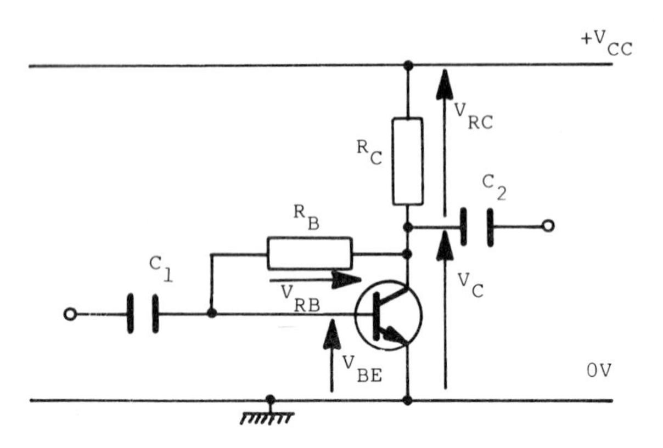

Fig. 3.5 Feedback bias stabilisation

The best and most common method of providing stabilisation is 'base-voltage bias'. Figure 3.6 shows the basic circuit. The transistor is connected in common-emitter mode. Capacitors C_{IN} and C_{OUT} are d.c.-blocking capacitors, to prevent the input and output circuits from affecting the d.c. bias conditions. Resistor R_C is the collector load resistor, to convert current changes into voltage. Resistors R_1 and R_2, in conjunction with R_E, provide the bias voltage for the transistor. Resistor R_E stabilises the operating point against changes caused by variations in temperature – affecting leakage current I_{CBO} – change of active device, ageing, etc.

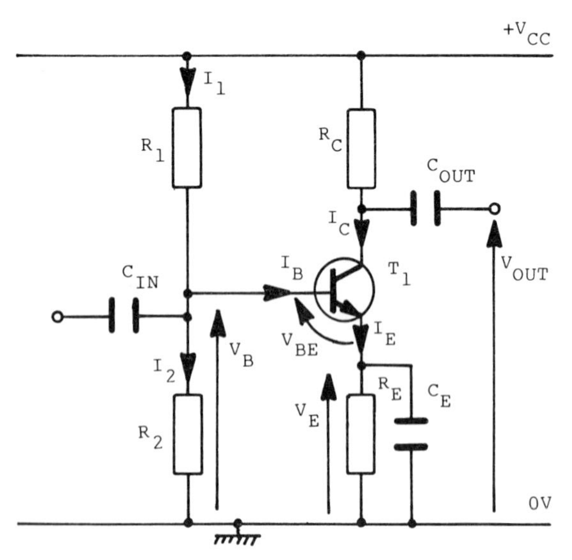

Fig. 3.6 Common-emitter base-voltage bias

The operation of the circuit is based on the assumption that V_B is for all practical purposes constant – this will be near enough the case if $I_1 \gg I_B$. If I_1 is made too large, it will be a drain on the power supply; $I_1 = 10I_B$ has been found to give satisfactory operation in most cases. V_E is usually chosen to be 10 to 15% of V_{CC}, typically 1 V. In class-A operation, I_{BQ} is set so that $I_{CQ} = h_{FE}I_{BQ}$ gives a collector voltage $V_C = V_{CC} - I_CR_C$ of approximately two thirds of the supply voltage. By Kirchoff's second law, $V_B = V_{BE} + V_E = V_{BE} + I_ER_E$. An increase in I_B, for any reason, will be amplified by normal transistor action, resulting in an increase in I_C and hence I_E (since $I_C \approx I_E$); thus $V_E = I_ER_E$ will increase. Since V_B is constant, $V_{BE} = V_B - V_E$ *must* fall. If V_{BE} falls, the input current I_B will be reduced and the conditions will revert almost to what they were before the initial increase in I_B took place. If I_B decreases, the reverse takes place. This stabilisation is achieved using negative feedback, covered in greater detail in chapter 7.

The effect of *any* current change in R_E is to alter the base–emitter voltage to try to prevent the current in R_E changing. If we do not wish the signal to be affected, we must arrange that any changes due to the signal do not affect the current in R_E. Capacitor C_E – the 'bypass' capacitor – effectively provides a short circuit across R_E to the signal. This is sometimes referred to as 'decoupling' the resistor.

3.3 The transistor as a switch

The simple transistor of fig. 3.7(a) can be used to represent a switch. With $V_{BE} = 0$, the output V_{CE} will be approximately V_{CC}: point A on fig. 3.7(b). If V_{BE} is increased sufficiently to ensure that I_B saturates the transistor, V_{CE} will fall quickly to $V_{CE(SAT)}$ (approximately zero): point B on fig. 3.7(b). Thus, if we arrange that the input voltage is either sufficient to saturate the transistor or small enough to ensure that the transistor is switched off, we can use the transistor as an electronic switch with two possible states. By

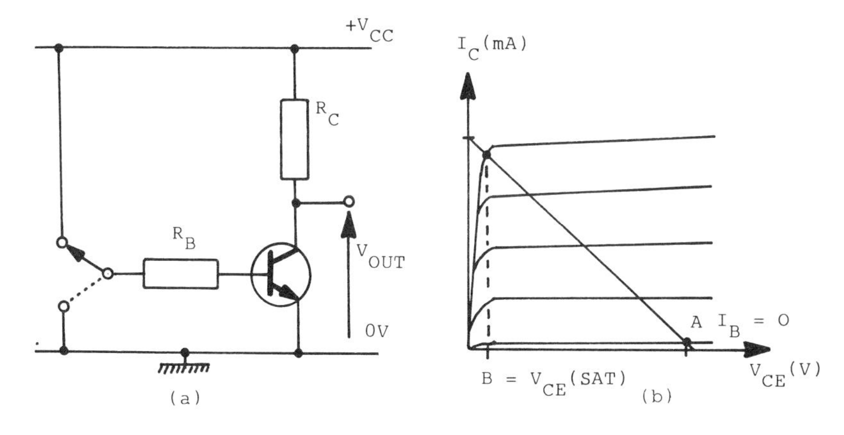

Fig. 3.7 Basic transistor switch

measuring (or detecting by some other means) the collector voltage, the 'state' of the switch can be determined and used.

The speed of transition between the two states can be made very fast – less than a microsecond. By suitable choice of transistor and resistors, the above operating conditions can be met, providing the basis of digital switching circuits.

3.4 The decibel

A very common and useful concept for measuring voltage, current, and power gains is the *decibel* (dB). The decibel – essentially a logarithmic power ratio – is one tenth of a *bel*, which is defined as the log to base 10 of the ratio of the two powers (output and input).

$$\text{In decibels,}\quad \text{power gain } A_{P(\text{dB})} = 10\log_{10}\frac{P_{\text{OUT}}}{P_{\text{IN}}}\ \text{dB}$$

$$= 10\log_{10}A_P\ \text{dB} \qquad 3.5$$

For voltage gain, since $P = V^2/R$, we get

$$\text{gain } A_{V(\text{dB})} = 10\log_{10}\frac{V_{\text{OUT}}^2/R}{V_{\text{IN}}^2/R}\ \text{dB}$$

$$= 10\log_{10}\frac{V_{\text{OUT}}^2}{V_{\text{IN}}^2}\ \text{dB}$$

$$= 10\log_{10}\left(\frac{V_{\text{OUT}}}{V_{\text{IN}}}\right)^2\ \text{dB}$$

$$= 20\log_{10}\frac{V_{\text{OUT}}}{V_{\text{IN}}}\ \text{dB}$$

$$= 20\log_{10}A_V\ \text{dB} \qquad 3.6$$

A similar exercise can show that current gain is given by

$$A_{I(\text{dB})} = 20\log_{10}\frac{I_{\text{OUT}}}{I_{\text{IN}}}\ \text{dB}$$

$$= 20\log_{10}A_I\ \text{dB} \qquad 3.7$$

Note: equations 3.6 and 3.7 are true only if the input and output resistances are equal; if they are not, then the complete formula must be used. The dB is a ratio, *not* an absolute unit. The quantities P_{OUT}, V_{OUT}, I_{OUT}, etc. may be replaced by their equivalent small-signal changes, i.e. ΔP_{OUT}, ΔV_{OUT}, etc.

Example 3.1 The transistor in the circuit of fig. 3.8 is a germanium device with $h_{\text{FE}} = 100$, $V_{\text{CC}} = 6$ V, and $R_{\text{C}} = 3.3$ kΩ.

a) Calculate a suitable value for the bias resistor R_{B} if the quiescent collector voltage is to be 2.7 V.
b) What power will the transistor dissipate under quiescent conditions?

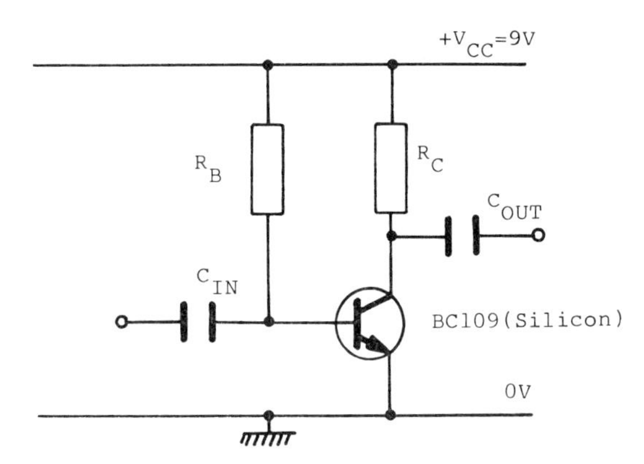

Fig. 3.8 Circuit for examples 3.1 and 3.2

a) If $V_{CQ} = 2.7$ V then $V_{RC} = V_{CC} - V_{CQ} = 6\text{ V} - 2.7\text{ V} = 3.3$ V

$$\therefore\ I_{CQ} = \frac{V_{RC}}{R_C} = \frac{3.3\text{ V}}{3.3\text{ k}\Omega} = 1\text{ mA}$$

$$I_{BQ} = \frac{I_{CQ}}{h_{FE}} = \frac{1\text{ mA}}{100} = 10\ \mu\text{A}$$

$$\therefore\ R_B = \frac{V_{CC} - V_{BE}}{I_{BQ}} = \frac{6\text{ V} - 0.2\text{ V}}{10\ \mu\text{A}} = 580\text{ k}\Omega$$

b) $P_{CQ} = V_{CQ} \times I_{CQ} = 2.7\text{ V} \times 1\text{ mA} = 2.7$ mW

Example 3.2 The transistor in the circuit of fig. 3.8 is a BC109, the supply voltage $V_{CC} = 9$ V, $R_C = 750\ \Omega$, and the output characteristics are as shown in fig. 3.9.

a) Draw a load line.
b) If the quiescent collector voltage is to be 5.5 V, calculate (i) a suitable value for the bias resistor R_B, (ii) the quiescent power dissipated by the transistor.
c) If the input signal is a sine wave of the form $i = 5 \sin\omega t\ \mu$A and the input resistance is 1.5 kΩ, calculate, in dB, (i) the current gain, (ii) the voltage gain, (iii) the power gain.
d) Calculate the r.m.s. value of the output current.

a) The load line is positioned using

$$V_C = V_{CC} = 9\text{ V} \quad \text{when} \quad I_C = 0$$

and $$I_C = \frac{V_{CC}}{R_C} = \frac{9\text{ V}}{750\ \Omega} = 12\text{ mA} \quad \text{when} \quad V_C = 0$$

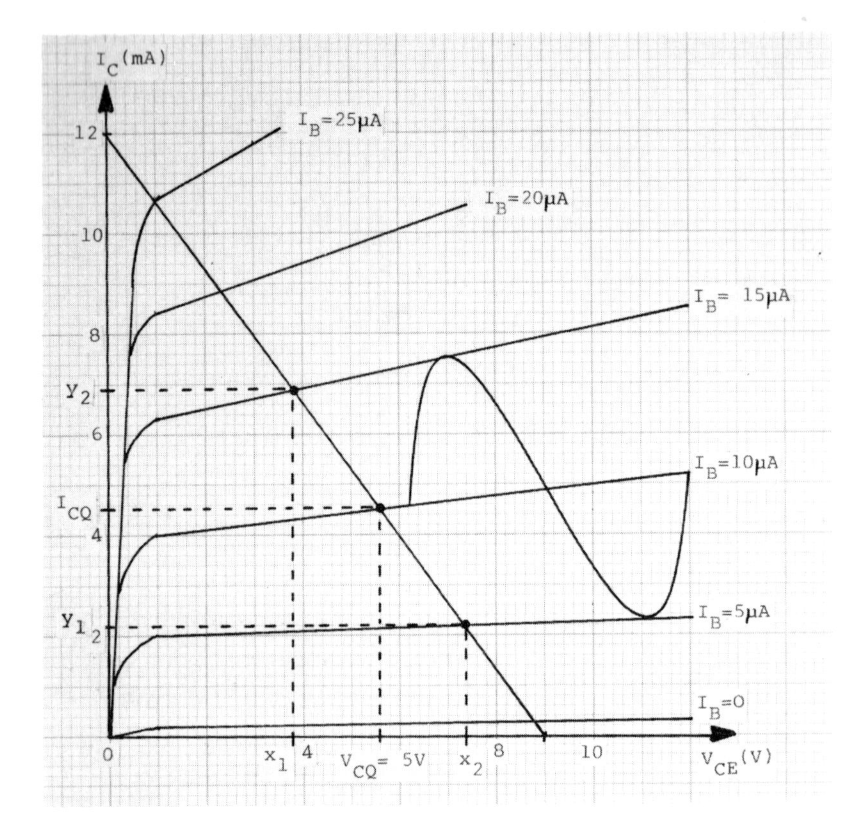

Fig. 3.9 Characteristics for example 3.2

b) (i) From $V_{CQ} = 5.6$ V then $I_{BQ} = 10\ \mu$A

$$\therefore \quad R_B = \frac{V_{CC} - V_{BE}}{I_{BQ}} = \frac{9\text{ V} - 0.6\text{ V}}{10\ \mu\text{A}} = 840\text{ k}\Omega$$

ii) $P_{CQ} = V_{CQ} \times I_{CQ} = 5.6\text{ V} \times 4.5\text{ mA} = 25.2\text{ mW}$

Note: $I_{CQ} = 4.5$ mA is obtained from the load line and characteristics of fig. 3.9.

c) (i) $\Delta I_C = y_2 - y_1 = 6.9\text{ mA} - 2.2\text{ mA} = 4.7\text{ mA}$

$$\Delta I_B = 15\ \mu\text{A} - 5\ \mu\text{A} = 10\ \mu\text{A}$$

$$\text{Current gain } A_{I(\text{dB})} = 20\log_{10}\frac{\Delta I_C}{\Delta I_B} = 20\log_{10}\frac{4.7\text{ mA}}{10\ \mu\text{A}}$$

$$= 20\log_{10} 470 = 53.4\text{ dB}$$

ii) $\Delta V_C = x_2 - x_1 = 7.4\text{ V} - 3.8\text{ V} = 3.6\text{ V}$

$\Delta V_B = \Delta I_B \times R_{IN} = 10\ \mu\text{A} \times 1.5\text{ k}\Omega = 15\text{ mV}$

$$\text{Voltage gain } A_{V(\text{dB})} = 20\log_{10}\frac{\Delta V_C}{\Delta V_B} = 20\log_{10}\frac{3.6\text{ V}}{15\text{ mV}}$$

$$= 20\log_{10}240 = 47.6\text{ dB}$$

iii) Power gain $A_{P(\text{dB})} = 10\log_{10}A_P$ dB

but $$A_P = \frac{\Delta P_{OUT}}{\Delta P_{IN}} = \frac{\Delta I_C{}^2 R_C}{\Delta I_B{}^2 R_{IN}} = \left(\frac{\Delta I_C}{\Delta I_B}\right)^2 \frac{R_C}{R_{IN}} = A_I^2\frac{R_C}{R_{IN}}$$

$$\therefore \quad A_{P(\text{dB})} = 10\log_{10}\frac{A_I^2 R_C}{R_{IN}} = 10\log_{10}110\,450 = 50.4\text{ dB}$$

d) R.M.S. value of $I_C = 0.707\ \Delta I_C/2 = 1.66\text{ mA}$

3.5 Amplifier classification and class of bias

Amplifiers may be classified in many ways; for example

a) by their frequency range – wideband or untuned, narrow or tuned, video or radio frequency, etc.;
b) by their use – audio, pulse, etc.;
c) by the method of operation – d.c., a.c., etc.;
d) by the method of interstage coupling – d.c., resistance–capacitance (*R–C*) coupling, transformer, etc.;
e) by the position of the quiescent or d.c. operating point.

It is (e) above which is used to define the '*class*' of operation – A, B, AB, or C. This usually done in terms of the relationship between the input signal and the load current. This is illustrated by the combined transfer characteristics shown in figs 3.10 to 3.13.

3.5.1 Class A

In class-A operation (fig. 3.10), the operating point and input signal are chosen so that current flows in the load for the whole of the input-current cycle. The amplifier normally operates over the linear part of its characteristics. Class-A operation is used for most 'small-signal' voltage amplifiers.

3.5.2 Class B

In class-B operation (fig. 3.11), the amplifier is biased to cut off ($V_{CQ} = V_{CC}$) so that quiescent power dissipation is small. Load current flows only during one half cycle of the input signal. Class-B operation is used for untuned power amplifiers such as audio-frequency power amplifiers.

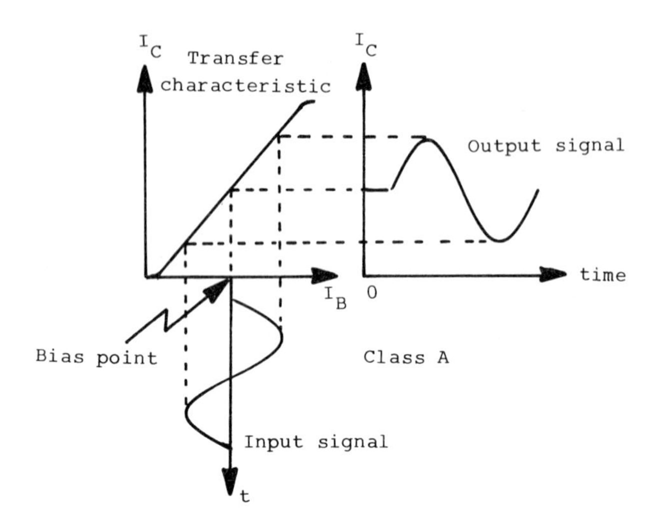

Fig. 3.10 Class-A bias

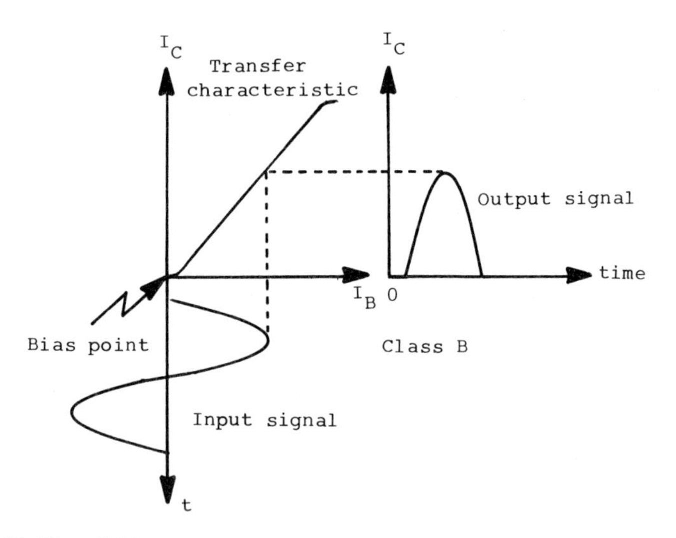

Fig. 3.11 Class-B bias

3.5.3 Class AB

This bias condition (fig. 3.12) is somewhere between the two previous classes. Class A is inefficient because power, $V_{CQ} \times I_{CQ}$, is being dissipated even when there is no signal applied. Class B introduces distortion due to the non-linearity of the characteristics close to cut-off. In class AB, the output current flows for more than half but less than the whole input signal. Its uses are as for class B.

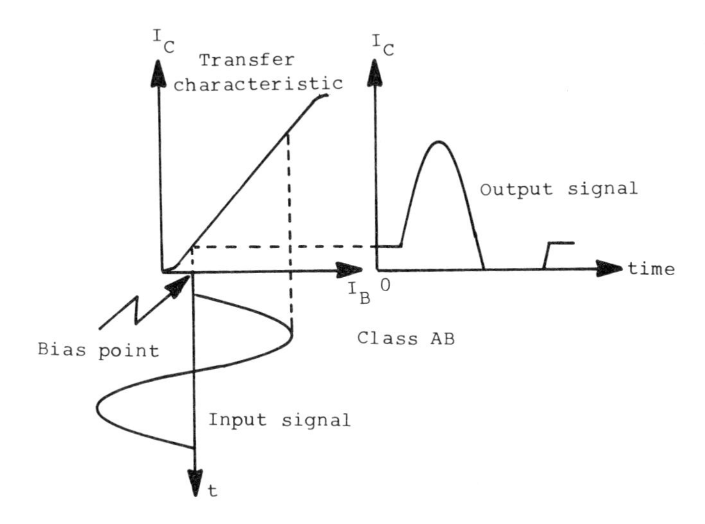

Fig. 3.12 Class-AB bias

3.5.4 Class C

The operating point in class-C operation (fig. 3.13) is chosen so that output current flows for less than a half cycle of the input signal. Class C is

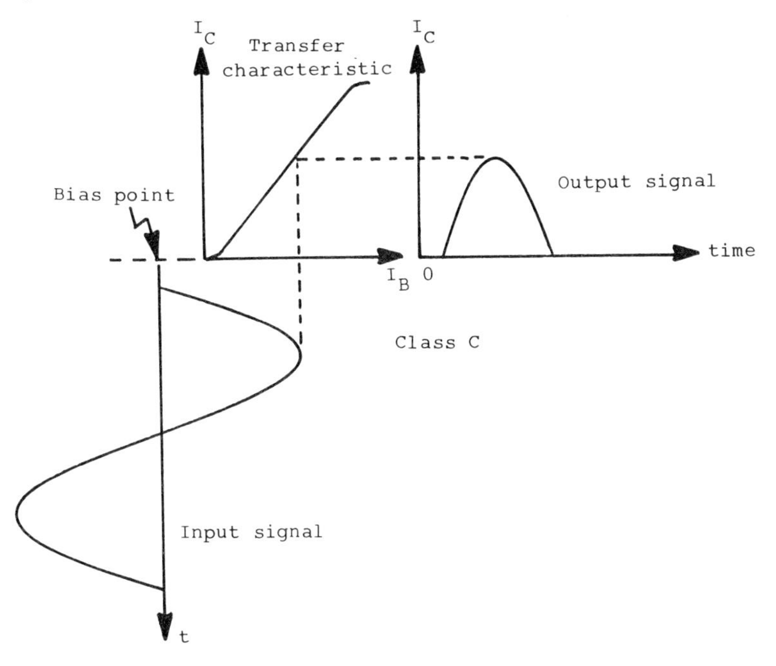

Fig. 3.13 Class-C bias

used for tuned radio-frequency amplifiers and oscillators and wave-shaping circuits.

The list of uses above is by no means exhaustive.

3.5.5 Two-stage class-A amplifiers

If the required gain cannot be achieved with a single stage, two or more stages may have to be used. The output of one stage must be connected to the input of the next stage. The most common method of connection – because it is cheap and introduces few problems – is 'resistance–capacitance' (*R–C*) coupling. Figure 3.14 shows a two-stage *R–C*-coupled amplifier using transistors with base-voltage feedback. Capacitor C_C and resistor R_4 'couple' the signal from the collector of T_1 to the base of T_2.

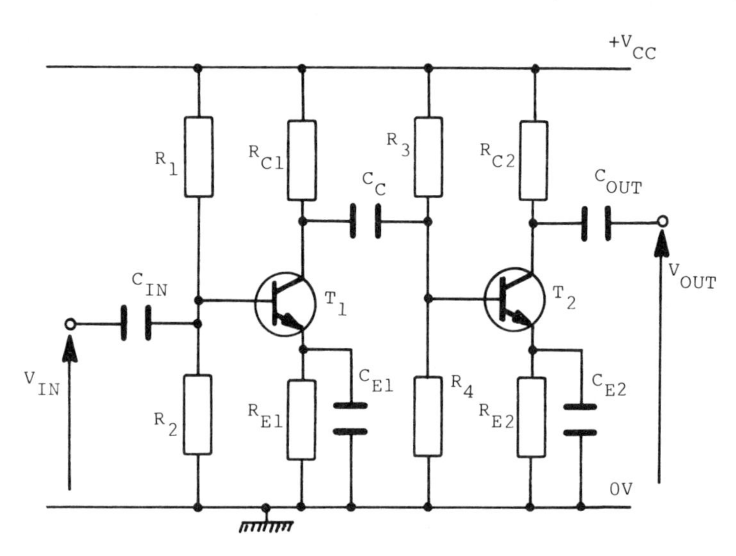

Fig. 3.14 Two-stage *R–C*-coupled amplifier

The transistors are connected in common-emitter mode. Resistors R_1 and R_2, in conjunction with R_{E1}, provide the bias voltage for transistor T_1. Resistor R_{E1} stabilises the operating point against changes caused by variations in temperature, ageing, etc. Resistor R_{C1} is the collector load and converts current changes into voltage. Capacitor C_{IN} is a d.c.-blocking capacitor to prevent the signal source affecting the biasing conditions of transistor T_1. Capacitor C_{E1} is used to 'bypass' or 'decouple' the emitter resistor, R_{E1}, for the a.c. signal. This is needed because the effect of R_{E1} is to try to stop *any* changes in the base–emitter voltage being amplified. It does this by feeding back any current changes in R_{E1} in opposition to the input signal. Because we wish to amplify the signal, we must arrange that any changes due to the signal do not affect the current in R_{E1}. Capacitor C_{E1} effectively provides a short circuit across R_{E1} for the signal.

Resistors R_3, R_4, R_{C2}, R_{E2} and capacitor C_{E2} carry out similar functions for transistor T_2 as the previous ones do for T_1. Note that R_4 has a dual function – as part of the bias for T_2 and as the 'R' part of the R–C coupling: the signal for the second stage is developed across R_4. Capacitor C_{OUT} prevents d.c. interaction between the collector circuit of T_2 and the external load.

3.6 Stage gain

Ideally an amplifier should have the same gain at any input frequency; in practice, this is not the case. If the gain of the amplifier in fig. 3.14 is plotted against a base of frequency, a graph similar to that in fig. 3.15 will result. Because of the large frequency range of this type of amplifier, a logarithmic scale is usually used for the frequency.

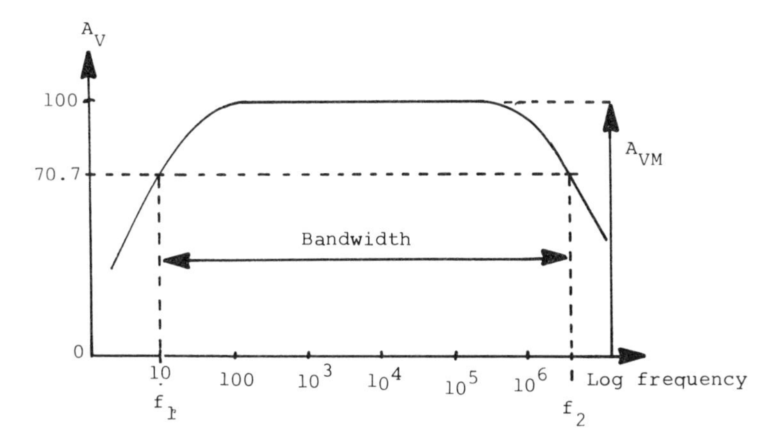

Fig. 3.15 Gain/frequency characteristic for the circuit of fig. 3.14

Over a large range of frequencies – approximately 100 Hz to 1 MHz in fig. 3.15 – the reactance of capacitors C_{IN}, C_C, C_{OUT}, C_{E1}, and C_{E2} is neglible and can be ignored and the gain is reasonably constant.

As the signal frequency falls, the reactance of these capacitors will increase. Thus part of the signal will be dropped across C_{IN}, C_C, and C_{OUT}, reducing the signal available to drive the transistor and thus reducing the overall gain. Also, because the reactance of the emitter-decoupling capacitor is increasing, more signal will be fed through the emitter resistors, causing negative feedback, as explained in section 3.2.2 for d.c. thermal stabilisation, and further reducing gain.

At high frequencies the gain also falls off, due to a number of factors. The connecting wire (or printed-circuit board (p.c.b.) track – this consists of copper strips or patterns which have been chemically produced on copper-covered sheets of insulating board) on both sides of capacitor C_C is separated from the ground rail by insulation, be it air or plastic. Since a capacitor is only two conductors separated by an insulating medium, the insulating connecting wires and ground rail can be considered as two capacitors connected as shown in fig. 3.16. The value of these 'stray capacitors' is only a few picofarads, but their effect become more marked as the frequency increases, effectively short-circuiting the signal to ground.

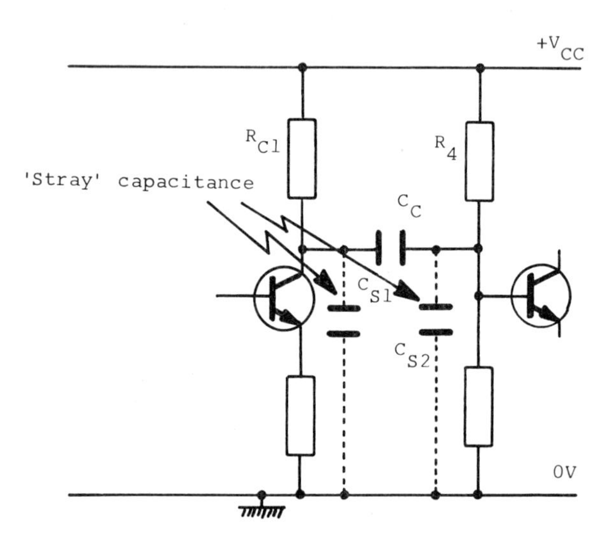

Fig. 3.16 Effective 'stray' capacitors

Two other factors can affect high-frequency gain: (i) the collector–base-junction capacitance of the transistor causes negative feedback and hence reduction of gain at very high frequencies; (ii) if the 'transit time' – the time an electron (or hole) takes to cross the base region of the transistor – is long compared to the period of the signal frequency, the gain will again be reduced.

The range of frequencies over which we consider an amplifier to be satisfactory is called the 'bandwidth'. It might seem a good idea to define bandwidth as that range of frequencies where the gain is constant; however, in the 'real' world we must accept some tolerance, and the generally accepted definition of bandwidth is 'that range of frequencies between the points where the voltage gain has fallen to 0.707 times the mid-band value' (i.e. to $0.707A_{VM}$). If the gain is in decibels, these points correspond to the points where the gain will have fallen by 3 dB from its mid-band value, the '3 dB points'. The choice of these points as the reference is arbitrary, but is a good round figure since it is where the power in the signal has fallen by a half.

Bandwidth is then equal to $f_2 - f_1$, which in most cases approximates to f_2.

3.7 Dynamic or a.c. load line

In practice, the load line is not the same for d.c. and a.c. (signal) conditions – this is because all capacitors and power supplies (including batteries) are for practical purposes short circuits to a.c. As a result, two load lines have to be drawn on the characteristics to enable gains to be determined. Both load lines *must* pass through the same quiescent point.

Consider the common-emitter amplifier of fig. 3.17 and the transistor output characteristics shown in fig. 3.18.

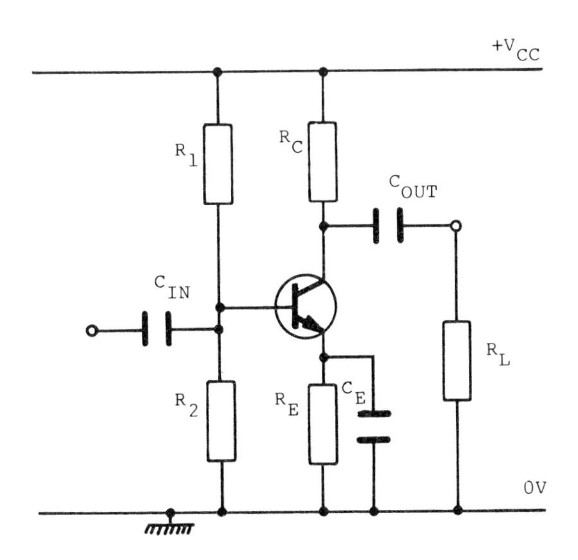

Fig. 3.17 Common-emitter amplifier

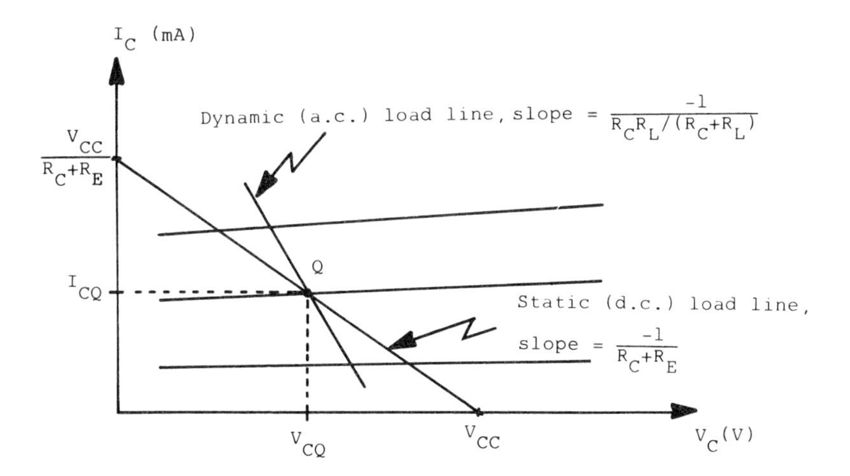

Fig. 3.18 Characteristics and load line for fig. 3.17

3.7.1 D.C. load line

To position the d.c. load line we need two points. As before, one point is found using

$$V_{CE} = V_{CC} \quad \text{when} \quad I_C = 0$$

When $V_{CE} = 0$, the supply voltage is then dropped across the emitter and collector resistors in series. For all practical purposes, the emitter current can be taken to be equal to the collector current, so we can locate the second point using

$$\text{when} \quad V_{CE} = 0 \quad \text{then} \quad I_C = \frac{V_{CC}}{R_E + R_C}$$

The load line can then be positioned and the quiescent point can be chosen.

3.7.2 A.C. load line

For a.c., resistor R_E is short-circuited by C_E and resistor R_L is now in parallel with R_C – since C_{OUT} and the power supply are effectively short-circuit. The equivalent circuit is shown in fig. 3.19.

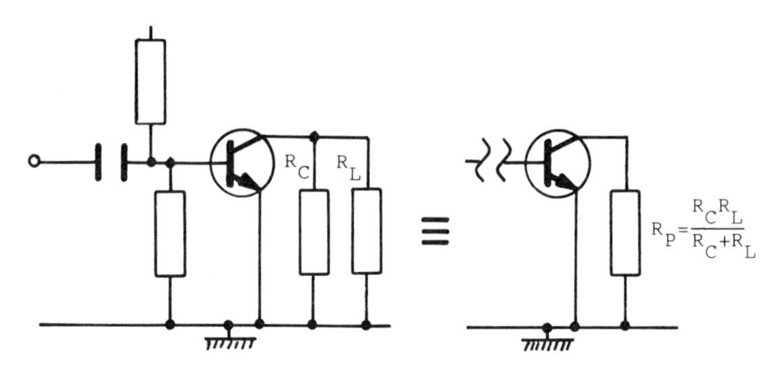

Fig. 3.19 Equivalent circuit for fig. 3.17

To position the a.c. load line we need two points. We have one – the quiescent point. To find another, we use the slope of the line, $-1/R_P$. For the circuit of fig. 3.19 'R_P' is the parallel combination of R_C and R_L.

Now the slope, $-1/R_P$, can also be found using $\Delta I_C/\Delta V_C$.

$$\therefore \quad \frac{\Delta I_C}{\Delta V_C} = \frac{-1}{R_P}$$

Since we know R_P, if we choose ΔV_C to be some convenient value we can then calculate ΔI_C.

To position the second point we drop a line vertically from the quiescent point by an amount equal to ΔI_C and then move horizontally to the right by an amount equal to ΔV_C. This gives the location of the second point. Figure 3.20 illustrates this.

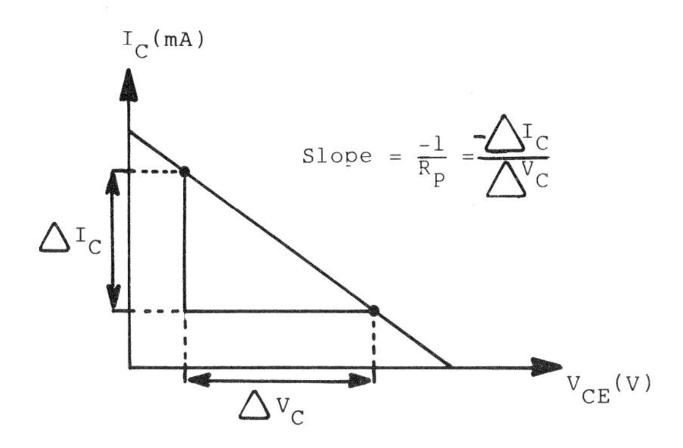

Fig. 3.20 Positioning the load line

Example 3.3 For the amplifier in fig. 3.17, $V_{CC} = 9$ V, $R_C = 680\ \Omega$, $R_E = 100\ \Omega$, and $R_L = 2$ kΩ. The transistor's characteristics are given in fig. 3.21.

a) Draw the d.c. load line and choose a suitable bias point for class-A operation.
b) Draw the a.c. load line.
c) For an input signal of $i = 5 \sin\omega t\ \mu$A, calculate the voltage and current gains if the input resistance is 2 kΩ.

a) The d.c. load line is positioned using

$$I_C = 0 \quad \text{when} \quad V_C = V_{CC} = 9\text{ V}$$

and when $\quad V_C = 0 \quad$ then $\quad I_C = \dfrac{V_{CC}}{R_C + R_E} = \dfrac{9\text{ V}}{780\ \Omega} = 11.5\text{ mA}$

A suitable bias point would be Q, with $I_B = 15\ \mu$A.

b) The a.c. load line is positioned using the slope, $-1/R_P$, where

$$R_P = \frac{R_C \times R_L}{R_C + R_L} = \frac{680\ \Omega \times 2000\ \Omega}{680\ \Omega + 2000\ \Omega} = 507\ \Omega$$

We now choose a convenient ΔV, say 2 V; then

$$\Delta I_C = \frac{\Delta V}{R_P} = \frac{2\text{ V}}{507\ \Omega} = 3.9\text{ mA}$$

Drop a vertical line 3.9 mA from point Q, then move horizontally to the right by 2 V to locate the second point. The load line can now be drawn.

c) To calculate voltage and current gains for an input signal of $i = 5 \sin\omega t\ \mu$A, we only need consider the peak values of the waveform, giving a $\Delta I_B = 10\ \mu$A. Thus

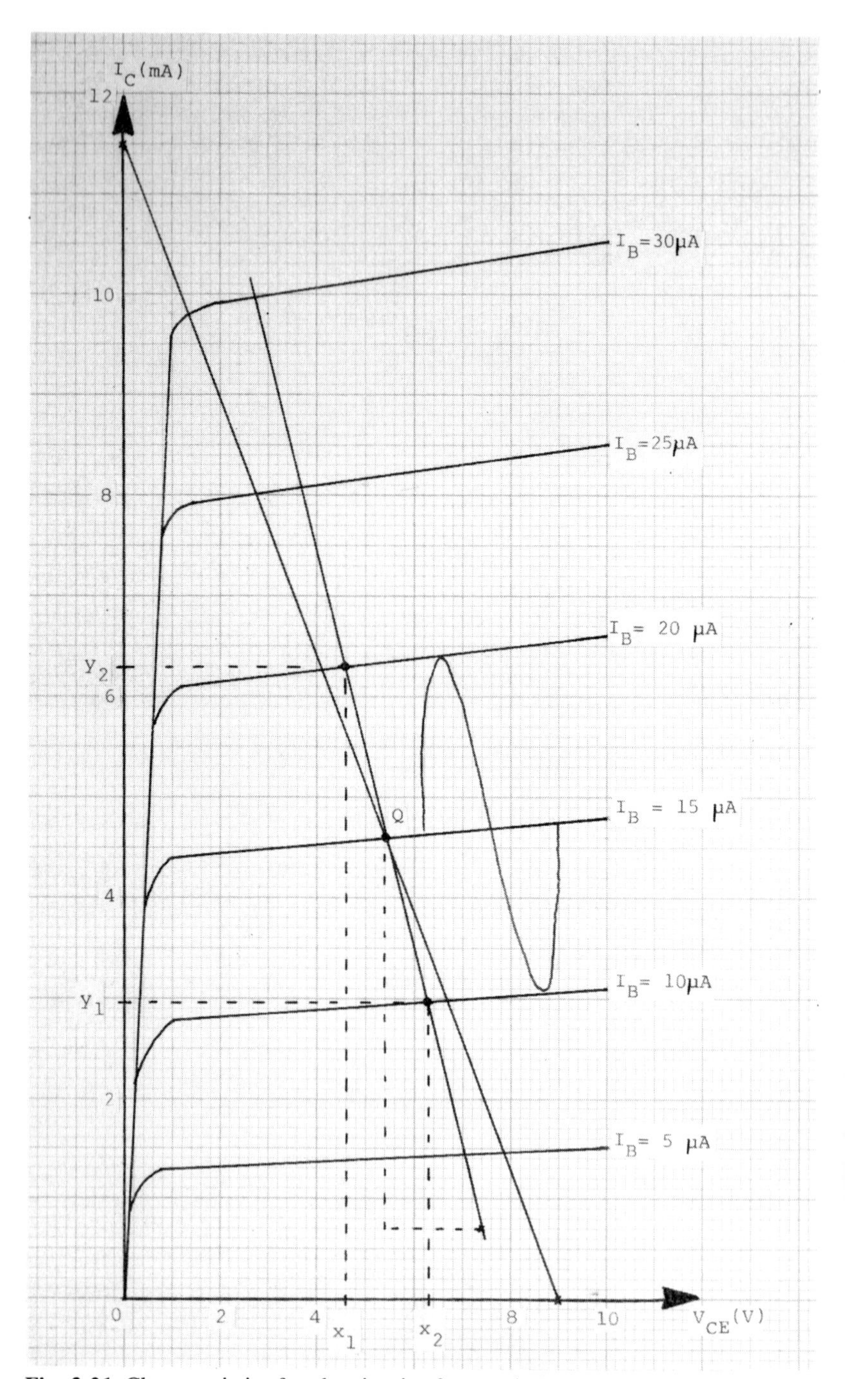

Fig. 3.21 Characteristics for the circuit of example 3.3

$$A_V = \frac{\Delta V_{OUT}}{\Delta V_{IN}} = \frac{\Delta V_{OUT}}{\Delta I_B R_{IN}}$$

$$= \frac{x_2 - x_1}{10\ \mu A \times 2\ k\Omega} = \frac{6.3\ V - 4.6\ V}{20\ mV}$$

$$= \frac{1.7\ V}{0.02\ V} = 85$$

$$A_I = \frac{\Delta I_C}{\Delta I_B} = \frac{y_2 - y_1}{\Delta I_B}$$

$$= \frac{6.3\ mA - 2.95\ mA}{10\ \mu A} = \frac{3.35\ mA}{10\ \mu A} = 335$$

The values of x_1, x_2, y_1, and y_2 are found by taking horizontal and vertical lines from the intersection of the a.c. load line *and* the $I_B = 20\ \mu A$ and $I_B = 10\ \mu A$ curves.

Problems

The characteristics in figs 3.23, 3.25, 3.26, 3.27, 3.28, 3.29, and 3.30 may be photocopied for use by students or teachers.

3.1 If the transistor of fig. 3.22 is a germanium device with $h_{FE} = 100$, $V_{CC} = 6$ V, and $R_C = 3.3$ kΩ, calculate the value of R_B to give $V_{CQ} = 2.7$ V.

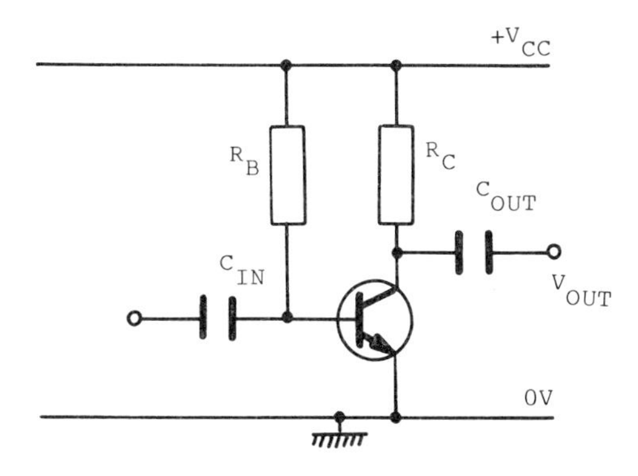

Fig. 3.22 Circuit for problem 3.1

3.2 The common-emitter amplifier in fig. 3.22 has a supply voltage $V_{CC} = 12$ V, $R_C = 200\ \Omega$, and the characteristics of fig. 3.23. Draw the d.c. load line and choose a suitable bias point. Hence calculate

a) a suitable value for R_B (the transistor is silicon),
b) the power dissipated by the transistor under quiescent conditions,
c) the current gain if the input signal is a sine wave of peak value 0.1 mA,
d) the voltage gain if the input resistance is 500 Ω,
e) the power gain.

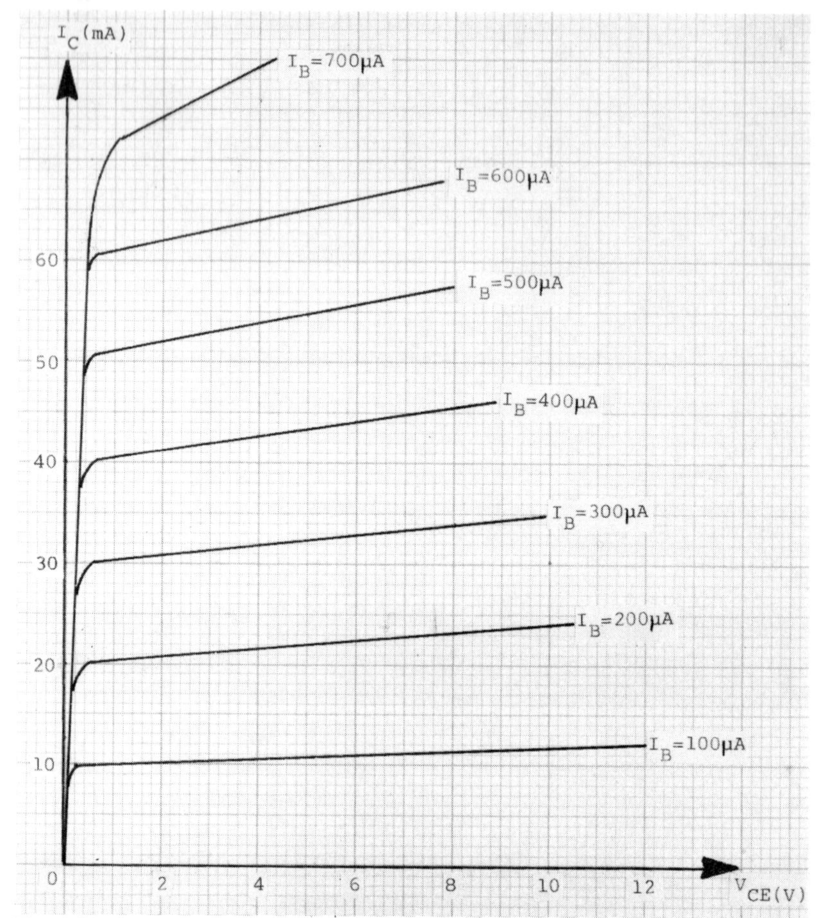

Fig. 3.23 Characteristics for circuit of problem 3.2

3.3 If current flows in the resistors R_B in figs 3.24(a) to (i) (apart from any leakage current, which can be neglected), the transistor *will* saturate. For each of the circuits, determine the reading on the voltmeter if V_{CC} = 10 V in each case.

3.4 The transistor in the amplifier of fig. 3.22 is a silicon type. V_{CC} = 10 V and R_C = 160 Ω. Figure 3.25 shows the output characteristics.

a) Draw a load line for the d.c. conditions, choose a suitable quiescent operating point, and calculate
   i) the value of the bias resistor R_B,
   ii) the p.d. across the collector load resistor R_C,
   iii) the current through the transistor,
   iv) the power dissipated by the transistor,
   v) the power dissipated by the load resistor R_C,
   vi) the power dissipated by the supply.

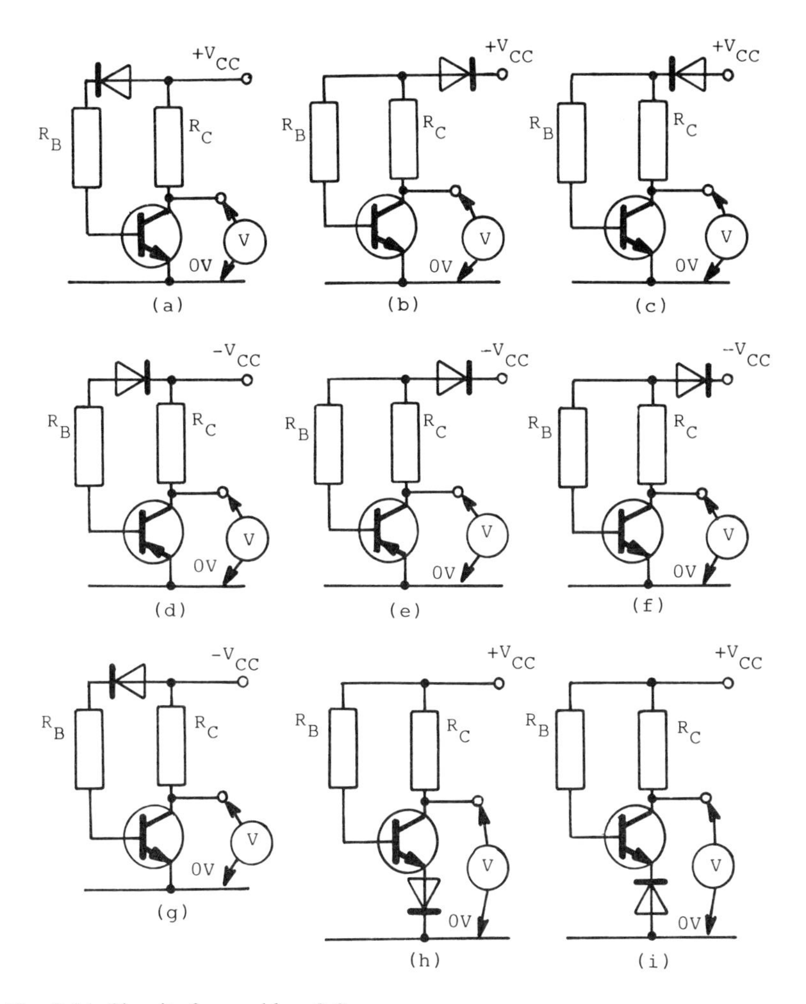

Fig. 3.24 Circuits for problem 3.3

b) If an input signal of 20 μA peak-to-peak value is applied and the input resistance is 1.2 kΩ, calculate
   i) the current gain,
   ii) the voltage gain in dB,
   iii) the power gain in dB.

3.5 For the circuit in fig. 3.22 and the output characteristics of fig. 3.26, if V_{CC} = 9 V and R_C = 750 Ω draw the d.c. load line.

a) Calculate a suitable value for R_B if the quiescent collector voltage is to be 5.5 V. (The transistor is silicon.)
b) For an input signal of 5 μA peak value and an input resistance of 1.5 kΩ, calculate, in dB, (i) the current gain, (ii) the voltage gain, (iii) the power gain.

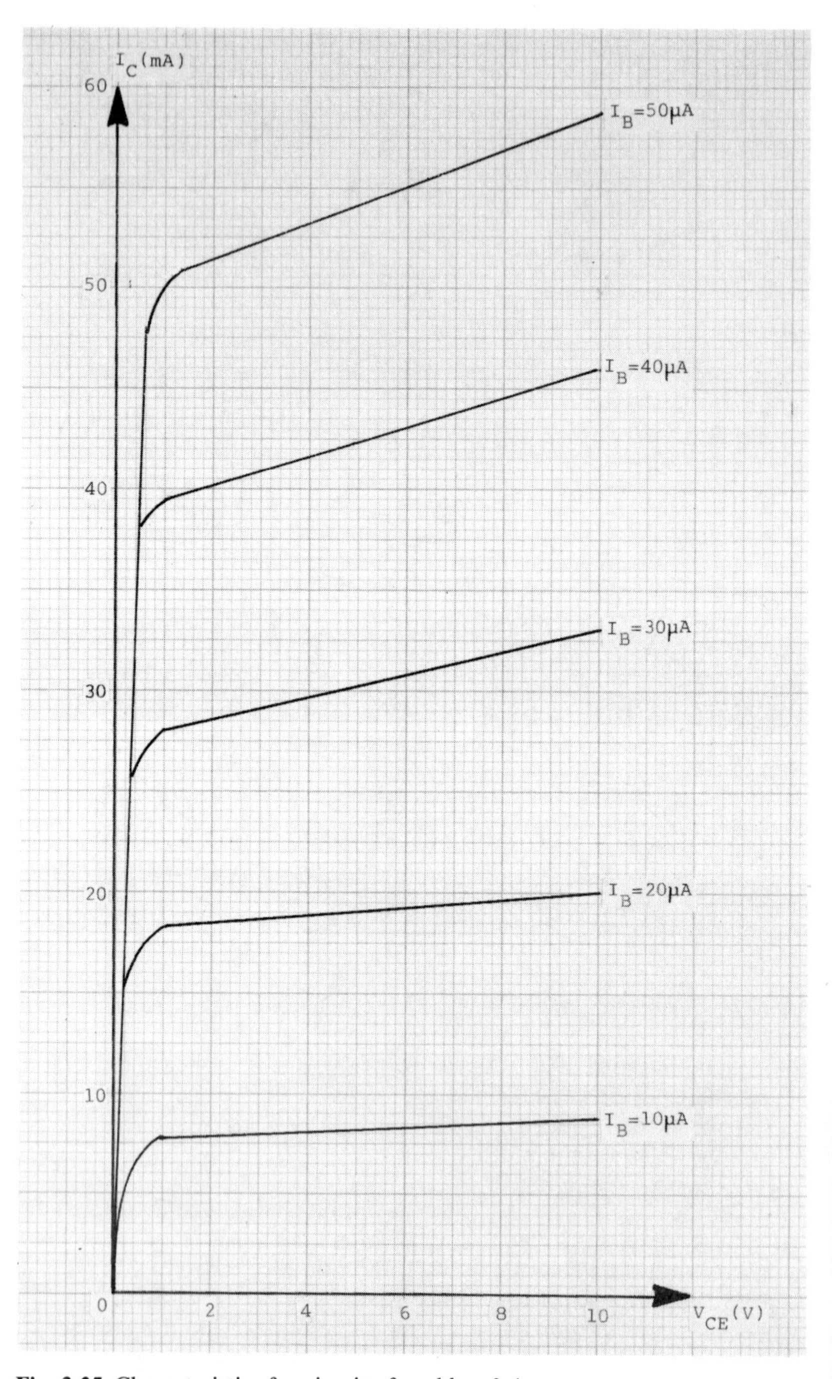

Fig. 3.25 Characteristics for circuit of problem 3.4

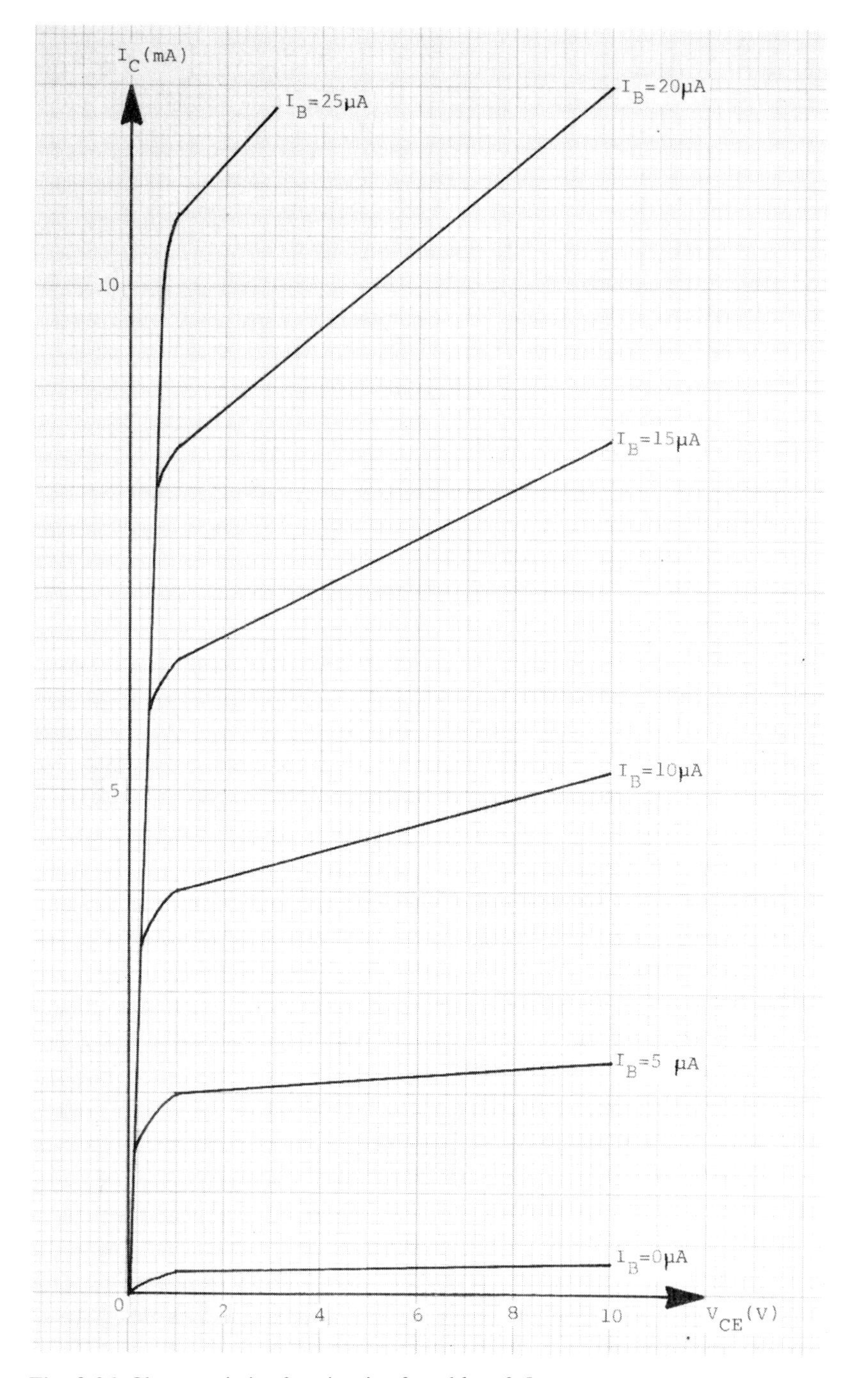

Fig. 3.26 Characteristics for circuit of problem 3.5

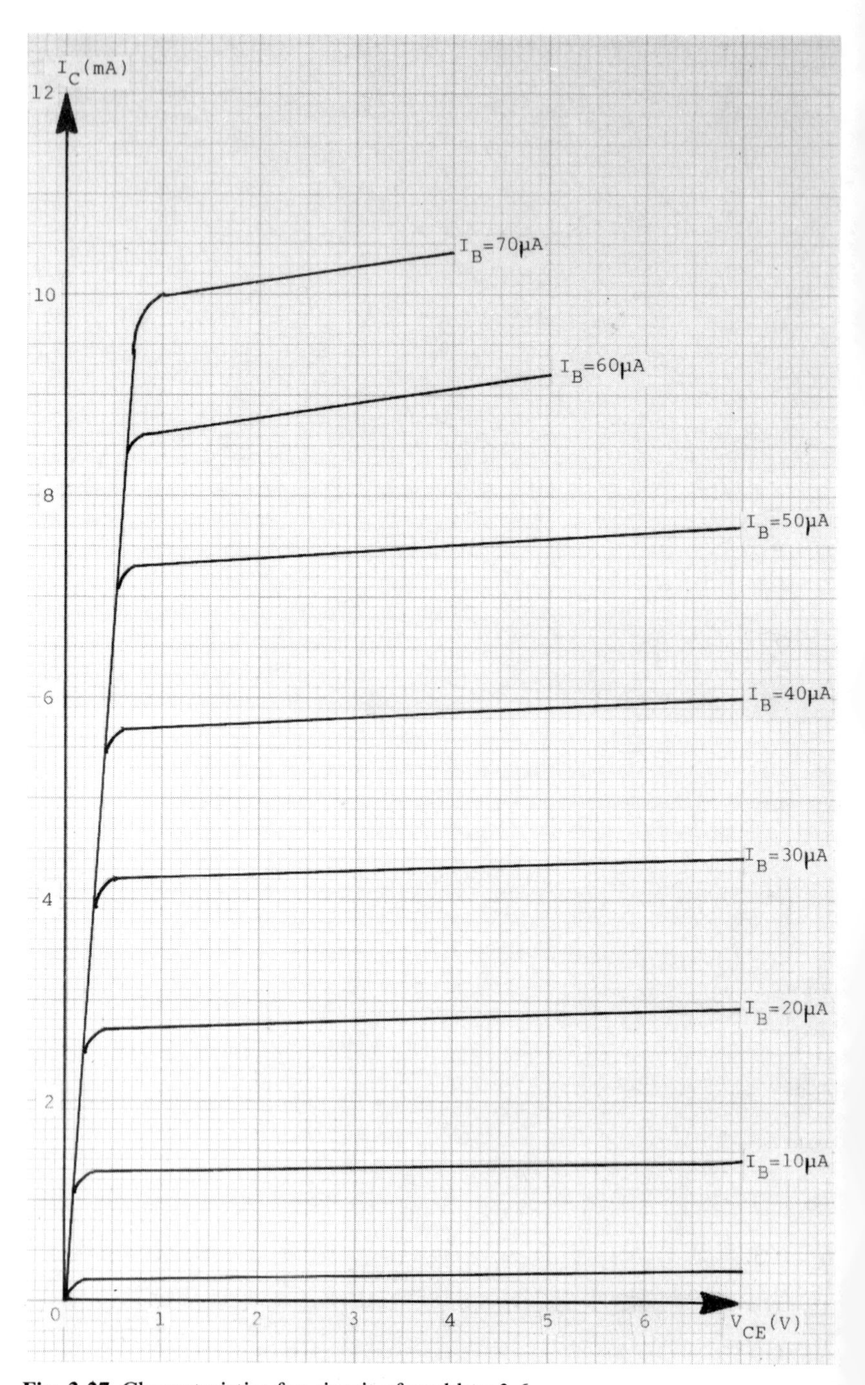

Fig. 3.27 Characteristics for circuit of problem 3.6

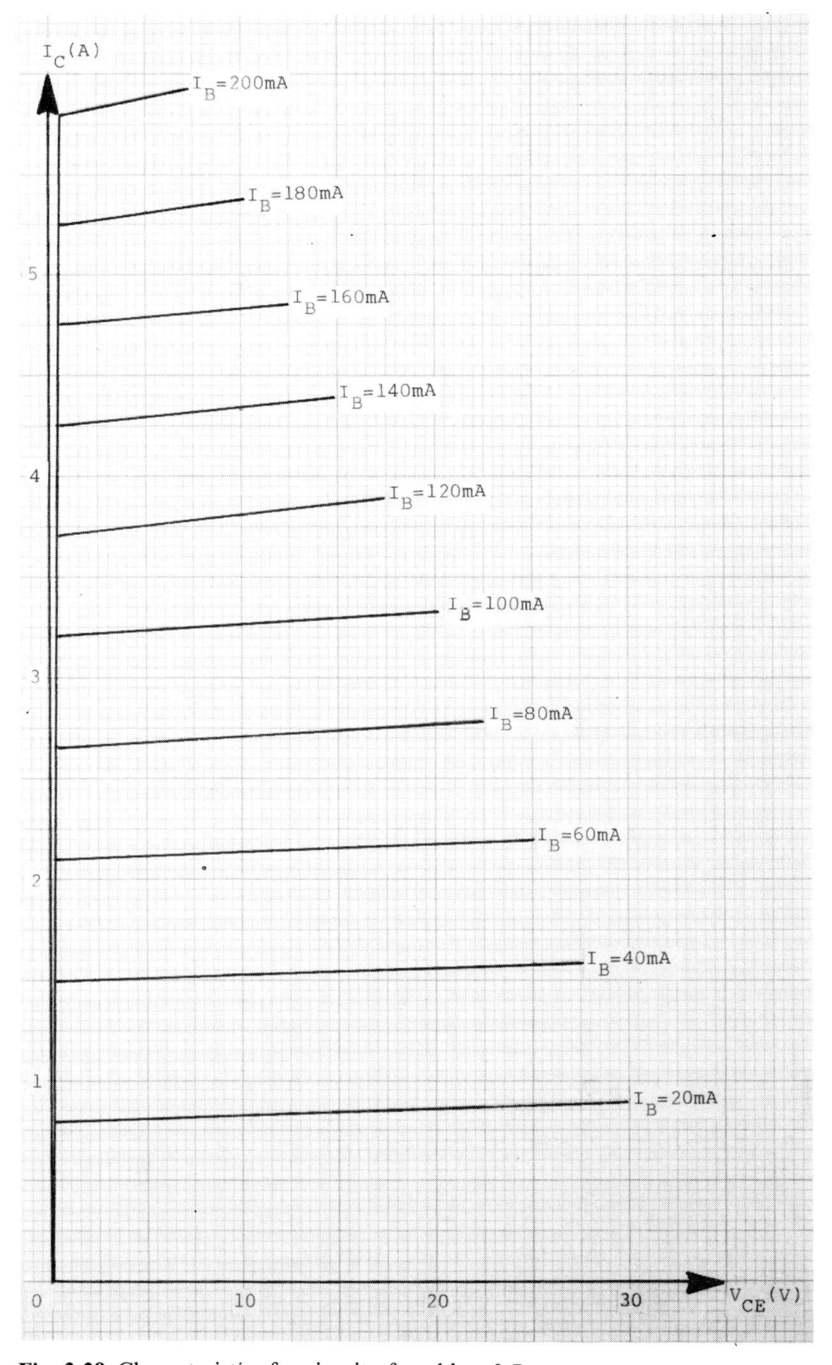

Fig. 3.28 Characteristics for circuit of problem 3.7

3.6 For the circuit of fig. 3.22, the characteristics of fig. 3.27, a supply voltage $V_{CC} = 6$ V, a quiescent collector current $I_{CQ} = 5.8$ mA, and a quiescent collector voltage $V_{CQ} = 3$ V,

a) draw the d.c. load line;
b) calculate suitable values for R_B and R_C;
c) if the input resistance of the circuit is 1.2 kΩ and an input signal of the form $i = 20 \sin\omega t\ \mu$A is applied, calculate in dB (i) the voltage gain, (ii) the current gain, (iii) the power gain.

3.7 Draw a load line on the characteristics of fig. 3.28 for the circuit of fig. 3.22 if $V_{CC} = 24$ V and $R_C = 4.8\ \Omega$. Choose a suitable bias point for an input signal of 60 mA peak value and calculate

a) a suitable value for R_B,
b) the quiescent power dissipated by the transistor,
c) h_{fe},
d) the power gain in dB if the input resistance is 120 Ω.

3.8 The circuit of fig. 3.17 is to be used as an amplifier. The characteristics of the transistor are given in fig. 3.29, $V_{CC} = 12$ V, $R_C = 120\ \Omega$, $R_E = 33\ \Omega$, and $R_L = 300\ \Omega$. Draw the d.c. load line and choose a suitable bias point. Hence construct the a.c. load line and use it to find

a) the current gain for an input signal $i = 50 \sin\omega t\ \mu$A,
b) the voltage gain for the same signal if the input resistance is 300 Ω.

3.9 A common-emitter transistor amplifier has the data given in Table 3.1. Plot the characteristics of the transistor. The transistor is to form a single-stage amplifier having an emitter resistor of 390 Ω and a collector load resistor of 2.2 kΩ. The supply voltage is 14 V, and an external load of 1830 Ω is connected to the output terminals.

a) Draw the d.c. load line and choose a suitable operating point.
b) Draw the a.c. load line.
c) Calculate the current gain when an input signal of 30 μA peak value is applied.

Table 3.1

	I_C (mA) for I_B (μA)		
V_{CE} (V)	20	50	80
2	1.7	3.2	4.9
4	1.9	3.6	5.3
6	2.1	3.8	5.5
8	2.2	4.0	5.7
10	2.4	4.2	5.9

3.10 The circuit of fig. 3.17 is to be used as an amplifier. The transistor is a BC109 whose characteristics are provided in fig. 3.26. $V_{CC} = 9$ V, $R_C = 820\ \Omega$, $R_E = 82\ \Omega$, and $R_L = 2$ kΩ.

a) Draw the d.c. load line and choose a suitable bias point.

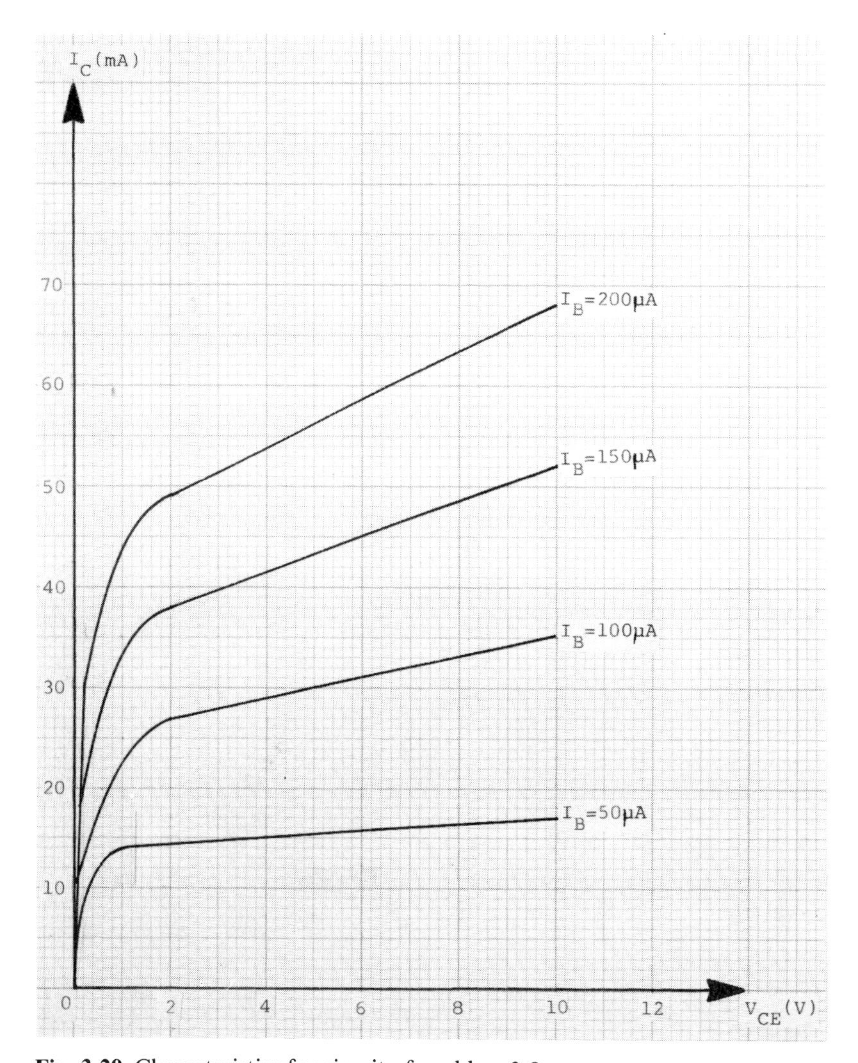

Fig. 3.29 Characteristics for circuit of problem 3.8

b) Draw the a.c. load line and, for an input signal $i = 5 \sin\omega t$ μA and an input resistance of 2 kΩ, calculate (i) the current gain, (ii) the voltage gain.

3.11 The characteristics in fig. 3.30 are for the transistor in fig. 3.17. $V_{CC} = 6$ V, $R_E = 56$ Ω, and $R_L = 1.8$ kΩ.

a) If the quiescent collector voltage is to be 3.5 V, draw the d.c. load line and use it to calculate the value of the collector load resistor R_C.

b) Draw the a.c. load line.

c) If the input resistance of the amplifier is 1.5 kΩ and an input signal of

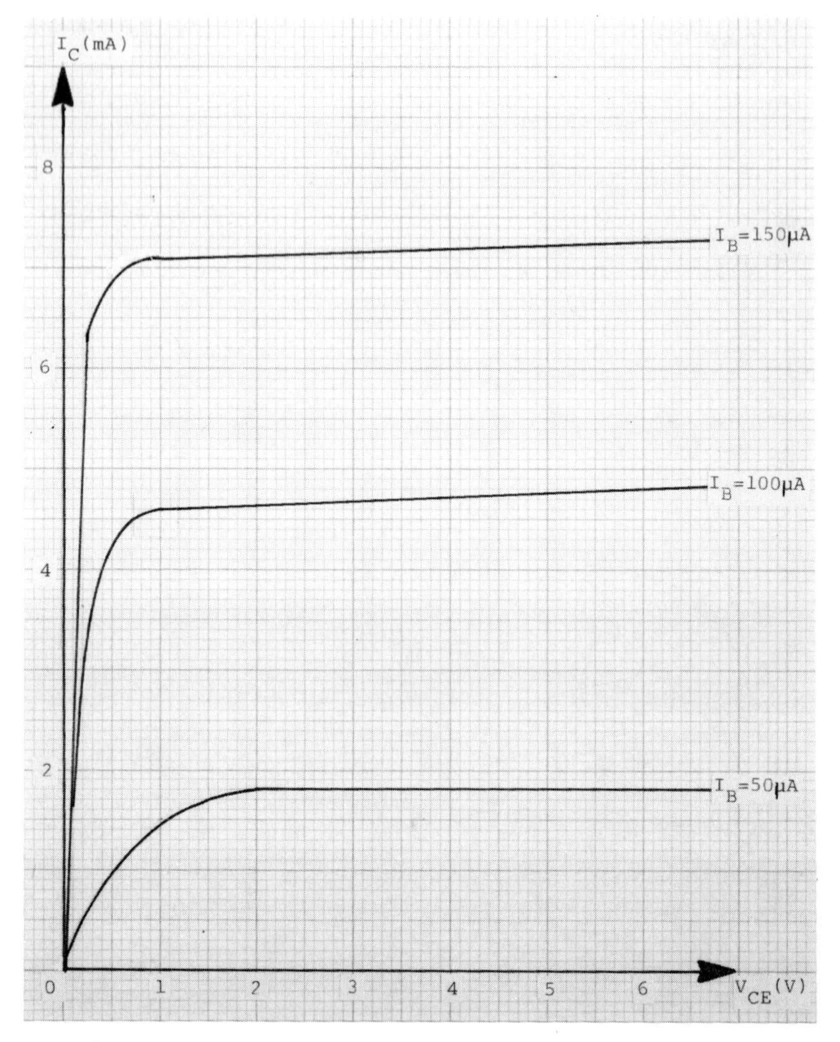

Fig. 3.30 Characteristics for problem 3.11

50 μA peak value is applied, calculate (i) the current gain, (ii) the voltage gain, (iii) the power gain.

3.12 What is the function of the collector load resistor in an amplifier?

3.13 When a common-emitter transistor is saturated, a further increase in base current causes

a) collector voltage to fall.
b) collector current to increase.
c) no change in collector voltage.
d) no change in collector current.

e) both (a) and (b).
f) both (c) and (d).
g) none of the above.

3.14 h_{fe} is generally slightly .... than h_{FE}.

3.15 Explain why the collector–emitter voltage of a saturated transistor can never fall to zero.

3.16 Explain the effects on its quiescent collector voltage of heating a transistor.

3.17 Explain the term 'thermal runaway' in relation to transistors.

3.18 If a common-emitter transistor amplifier is saturated and a sine-wave input signal is applied to the base, sketch one cycle of the input and output waveforms on the same time axis.

3.19 What is meant by 'small signal' in relation to transistor amplifiers?

3.20 Give reasons for the shape of the gain/frequency characteristic of a common-emitter amplifier.

3.21 What practical changes would be necessary to reduce the frequency of the lower 3 dB point of an amplifier?

4 Field-effect transistors

4.1 Introduction

Field-effect transistors (FET's) combine some of the advantages of the bipolar transistor and the thermionic vacuum valve, e.g. they are small like the bipolar transistor and have high input impedance like the vacuum valve. FET's are 'unipolar' devices because they operate using either holes or electrons – not both – as the current carriers. They have three terminals – called drain, source, and gate.

There are two kinds of FET: the junction-gate field-effect transistor – the JUGFET or JFET – and the metal–oxide–semiconductor field-effect transistor – the MOSFET. The FET has similar applications to that of the bipolar transistor, e.g. amplifiers, but the major impact of FET technology is in the construction of low-power high-density devices such as memory 'chips' for computers – these devices may use MOS or CMOS (a combination of p-channel and n-channel MOSFET's – 'complementary MOS') technology.

4.2 The junction-gate field-effect transistor (JUGFET or JFET)

The JUGFET consists of a 'channel', usually formed from n-type silicon (though p-type silicon may be used), with two p-type regions diffused on to the side to produce p–n junctions. Ohmic contacts (ohmic contacts do not involve a p–n junction) are formed to make the electrical connections for the terminals. We shall consider only the n-type JUGFET.

The construction is shown in fig. 4.1. A p-type substrate has an n-type layer diffused into it followed by another p-type layer, to form an 'n-channel' setting up two depletion layers as shown in fig. 4.1. Suitable ohmic aluminium contacts are added to make the drain, source, and gate terminals, and the surface is then protected by a layer of silicon dioxide.

Operation is easier to visualise if the device is considered to be a bar of n-type material with two p-type regions formed on the side, as in fig. 4.2. The resistance of the channel between drain and source can be increased by applying a reverse bias across the gate–source junctions on the sides of the bar – the bias expands the depletion regions towards the centre of the bar, effectively reducing the channel width and increasing its resistance.

If a voltage is now applied between drain and source, a 'drain current' I_D will flow. If an a.c. signal is now superimposed on to the gate bias, the width of the channel is controlled and the signal waveform will be impressed on to the drain current. If this current passes through an external load resistor, a voltage gain may be achieved.

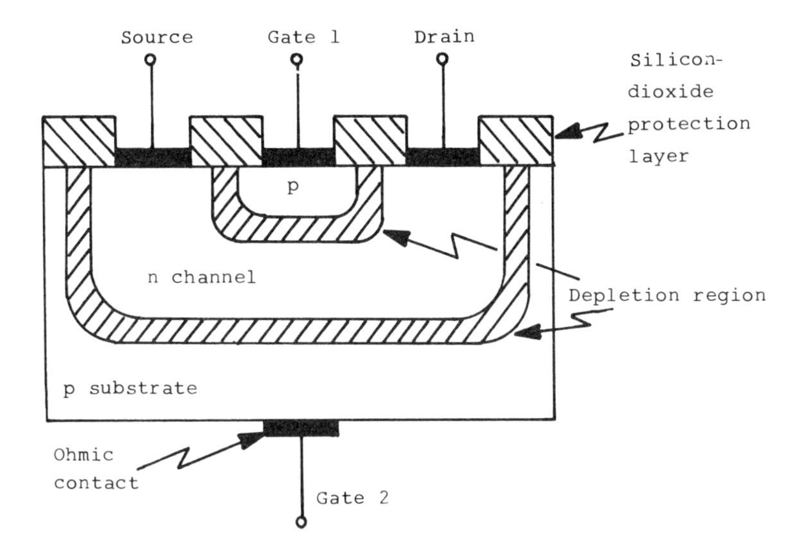

Fig. 4.1 Basic construction of a JUGFET

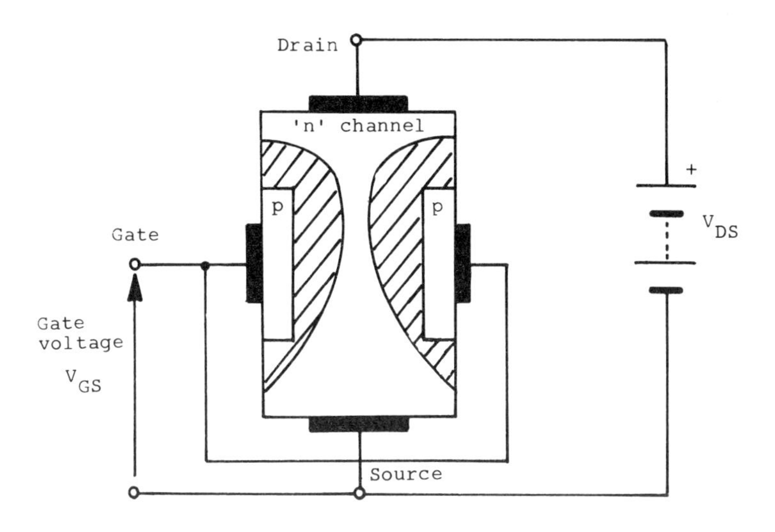

Fig. 4.2 FET operation

As the channel has a finite resistance, there will be a volt drop (which is approximately linear) between drain and source. The effect of this volt drop is to increase the reverse bias of the junction at the end nearest to the drain – this results in a 'wedge-shaped' channel as shown in fig. 4.2. If the drain voltage V_{DS} is increased, the depletion layers will eventually meet at the drain end of the channel. The channel is then said to be 'pinched-off' and the value of V_{DS} to achieve this is the 'pinch-off voltage', V_P. Further increase in V_{DS} produces little increase in drain current, as the depletion layer nearer the source end of the bar moves in. The value of V_P is less for higher values of reverse bias but is generally quoted for $V_{GS} = 0$ V.

The region of the characteristics beyond 'pinch-off' is known as the 'pinch-off region'. The device is normally operated in this region.

Figure 4.3 shows typical transfer (mutual) and drain (output) characteristics for a BFW10 JUGFET. The symbols for n-channel and p-channel JUGFET's are shown in fig. 4.4.

The input resistance of the JUGFET is that of a reverse-biased p–n junction and can be as high as 10^{10} Ω.

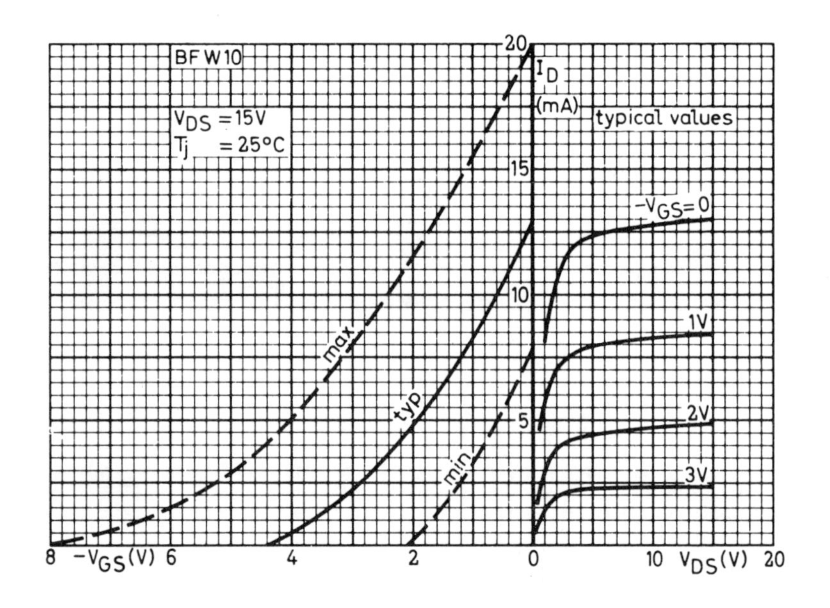

Fig. 4.3 Transfer (mutual) and drain characteristics for a BFW10

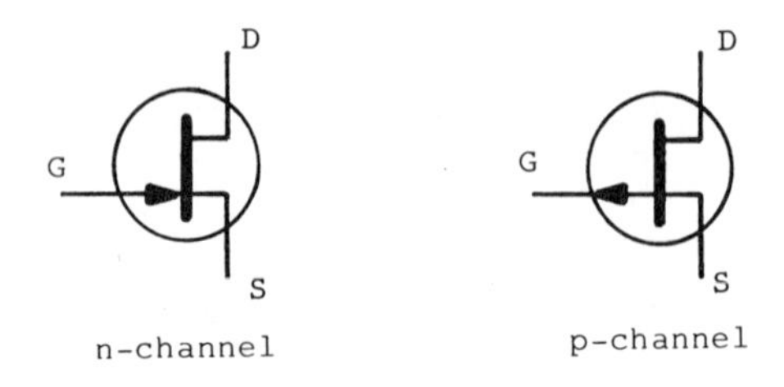

Fig. 4.4 JUGFET symbols

4.2.1 Device parameters

As with bipolar transistors, parameters are used to give a mathematical representation of the JUGFET. The most useful of these are the mutual conductance g_m (modern terminology uses 'transconductance', g_{fs}) and drain–source resistance r_{ds}.

4.2.2 Mutual conductance

The mutual conductance, g_m, is defined as the ratio of the small change in drain current to the small change in the gate–source voltage producing it, with the drain–source voltage constant. It is a measure of the effectiveness of a change in the input voltage, and its unit is the siemen (S). Mathematically, g_m is given by

$$g_m = \left.\frac{\Delta I_D}{\Delta V_{GS}}\right|_{V_{DS} = \text{constant}} \qquad 4.1$$

4.2.3 Drain–source resistance

The drain–source resistance, r_{ds}, is défined as the ratio of a small change in drain–source voltage to the small change in drain current producing it, with the gate–source voltage constant. This is the output resistance of the device, and its unit is the ohm. Mathematically, r_{ds} is given by

$$r_{ds} = \left.\frac{\Delta V_{DS}}{\Delta I_D}\right|_{V_{GS} = \text{constant}} \qquad 4.2$$

g_m and r_{ds} can be obtained from both transfer and output characteristics.

Typical parameter values

Drain-to-gate breakdown voltage	25 V
Mean operating drain current	5 mA
Mutual conductance	2 mS
Static drain–source resistance	1 kΩ

Example 4.1 Use equations 4.1 and 4.2 to obtain values for g_m and r_{ds} from (a) fig. 4.5(a), (b) fig. 4.5(b).

a) No values for V_{DS} and V_{GS} are given at which to obtain the parameters g_m and r_{ds} so we will use $V_{DS} = 20$ V and $V_{GS} = -0.5$ V.

From fig. 4.5(a),

$$g_m = \left.\frac{\Delta I_{D3}}{\Delta V_{GS2}}\right|_{V_{DS} = 20\text{ V}} = \frac{4.3\text{ mA} - 0}{1.08\text{ V} - 0}$$

$$= \frac{4.3\text{ mA}}{1.08\text{ V}} = 3.98\text{ mS}$$

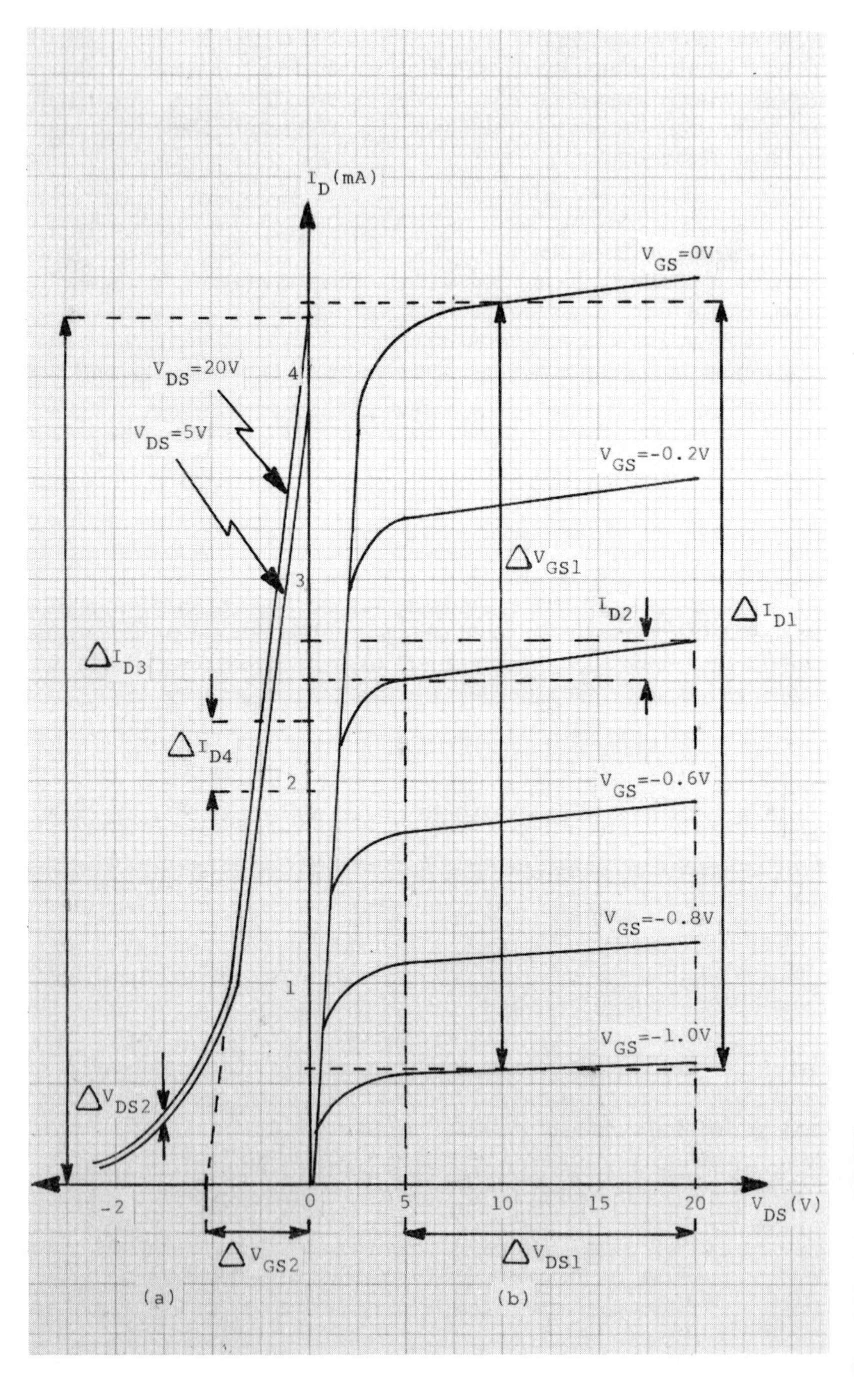

Fig. 4.5 Transfer (mutual) and drain characteristics for a BF256A

$$r_{ds} = \left.\frac{\Delta V_{DS2}}{\Delta I_{D4}}\right|_{V_{GS} = -0.5\ V} = \frac{20\ V - 5\ V}{2.30\ mA - 1.95\ mA}$$

$$= \frac{15\ V}{1.05\ mA} = 43\ k\Omega$$

b) Again, since no values for the constants are given we can choose our own, say $V_{DS} = 10$ V and $V_{GS} = -0.4$ V.

From fig. 4.5(b),

$$g_m = \left.\frac{\Delta I_{D1}}{\Delta V_{GS1}}\right|_{V_{DS} = 10\ V} = \frac{4.375\ mA - 0.575\ mA}{1\ V - 0}$$

$$= \frac{3.8\ mA}{1\ V} = 3.8\ mS$$

$$r_{ds} = \left.\frac{\Delta V_{DS1}}{\Delta I_{D2}}\right|_{V_{GS} = -0.4\ V} = \frac{20\ V - 5\ V}{2.7\ mA - 2.5\ mA}$$

$$= \frac{15\ V}{0.2\ mA} = 75\ k\Omega$$

Note that the values for g_m and r_{ds} are not the same in the two cases. This is due partly to the drawing of tangents to the curves, which eliminates some non-linearity errors but gives different values for the parameters; also, in the case of the mutual characteristics the actual curves are in practice much closer together. However it does emphasise that these quantities are *not* constants but depend upon the conditions under which they were obtained.

4.3 Metal-oxide-semiconductor field-effect transistors (MOSFET's)

(also known as insulated-gate field-effect transistors – IGFET's)

One technique used to make FET's is metal-oxide-semiconductor technology, in which an electric field is created in a metal gate which is insulated from the channel by an oxide layer. This technique has its major application in the construction of integrated circuits, particularly digital memory chips.

There are two types of MOSFET: the 'enhancement' and the 'depletion' types.

4.3.1 Enhancement MOSFET

Figure 4.6 shows the basic construction of an enhancement MOSFET. A p-type silicon substrate has two heavily doped (n^+ = heavily doped) regions with ohmic aluminium contacts for the drain and source terminals. A very thin layer (usually $< 1\ \mu m$) of silicon dioxide is then allowed to form before the gate connection is added. Since silicon dioxide is an extremely good insulator, the gate–source resistance of MOS devices is very high – up to $10^{15}\ \Omega$.

If a positive voltage V_{GS} is applied to the gate, it will induce a negative charge in the silicon dioxide next to the gate terminal; this results in a

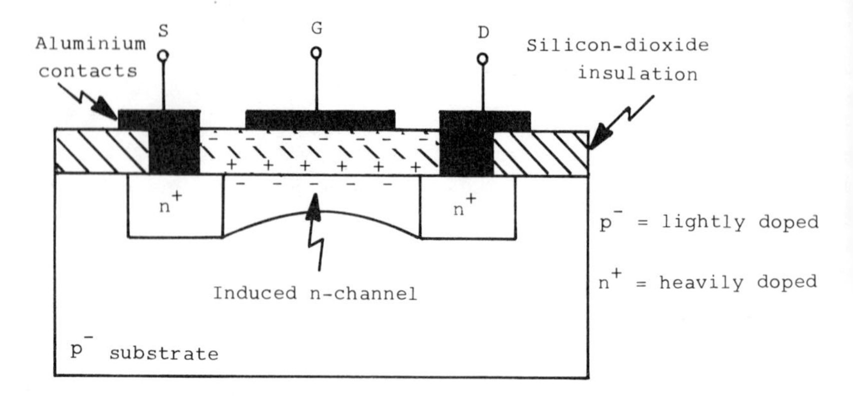

Fig. 4.6 Construction of an enhancement MOSFET

positive charge being induced in the silicon dioxide next to the substrate. This positive charge repels the majority carriers (holes) in the p-type substrate and attracts electrons, causing a build-up of electrons close to the surface and forming a negative channel between the heavily doped n^+ regions, permitting current to flow easily between drain and source. Figure 4.6 shows these effects.

If a positive voltage is now applied to the drain, electrons can flow from source to drain. Increasing V_{GS} widens or 'enhances' the channel; hence the name 'enhancement-type' MOSFET.

The value of V_{GS} at which the drain current falls to zero is referred to as the '*threshold voltage*', V_T, and it is analogous to pinch-off in the JUGFET.

Typical drain and mutual (or gate) characteristics are shown in fig. 4.7. Note in the case of the mutual characteristics that the gate is never made negative.

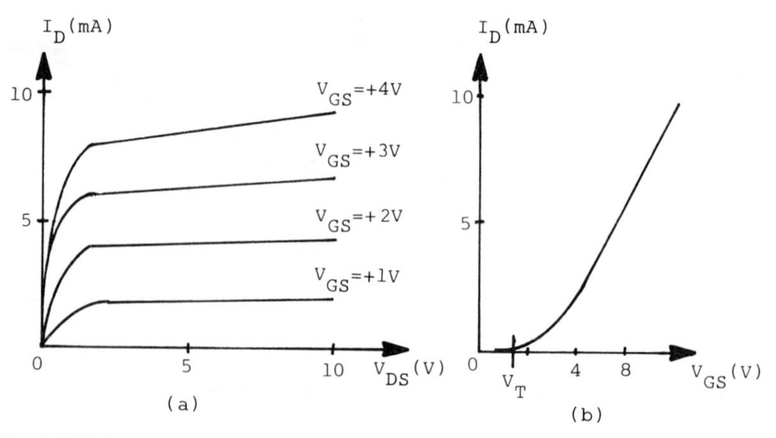

Fig. 4.7 Enhancement-MOSFET characteristics

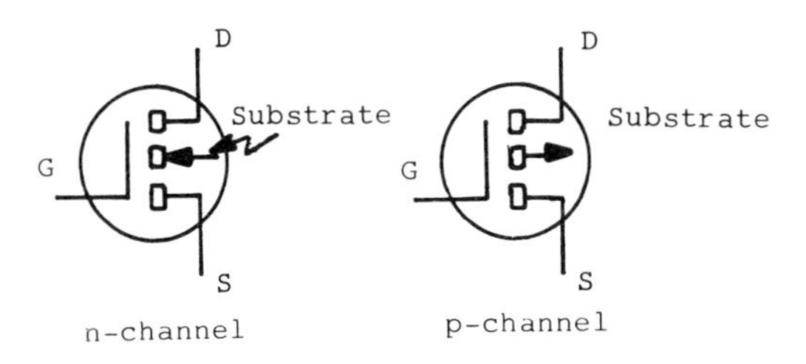

Fig. 4.8 Enhancement-MOSFET symbols

Figure 4.8 shows the symbols for n- and p-channel MOSFET's. Because these devices are normally open circuit, broken lines are used on the symbol. The arrow indicates polarity. The substrate is always shown, even though *no* physical connection can be made to it.

4.3.2 Depletion MOSFET

Figure 4.9 shows the construction of a depletion MOSFET. This differs from the enhancement MOSFET in that a moderately doped n-channel is introduced between the heavily doped drain and source regions. Thus, even without gate-to-source bias, a current will flow from drain to source if a voltage is applied to drain and source. If the gate is now made negative, the resultant electric field will force electrons out of the channel, i.e. 'depleting' the channel of charge carriers, so any current flow will be reduced. If a positive voltage is applied to the gate, the channel will be enhanced and current flow will increase. Thus this type of device can operate in both enhancement and depletion modes. However, its characteristics, shown in fig. 4.10, are not quite as linear as those of the enhancement type.

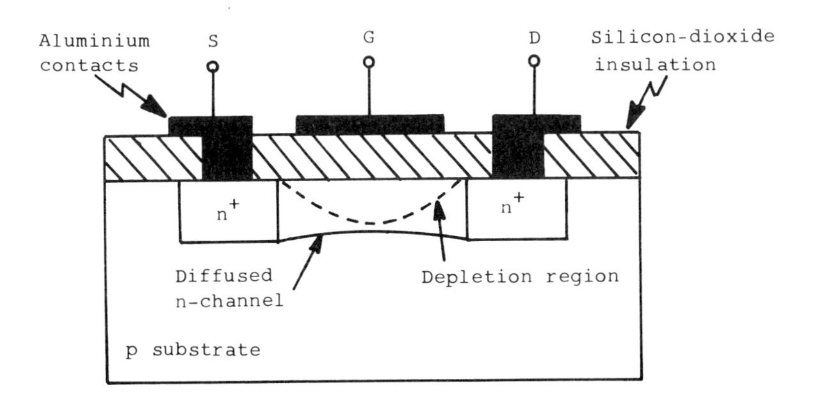

Fig. 4.9 Construction of a depletion MOSFET

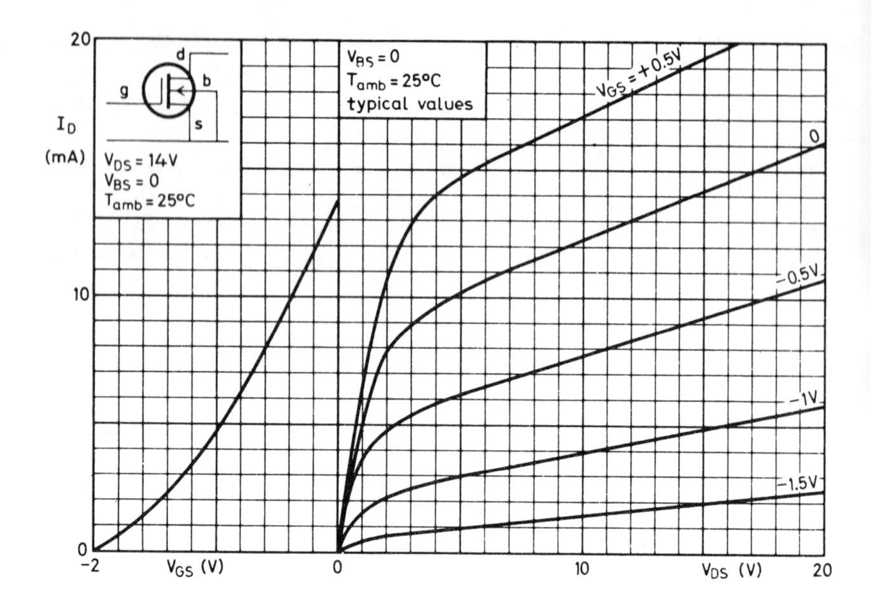

Fig. 4.10 Depletion-MOSFET characteristics

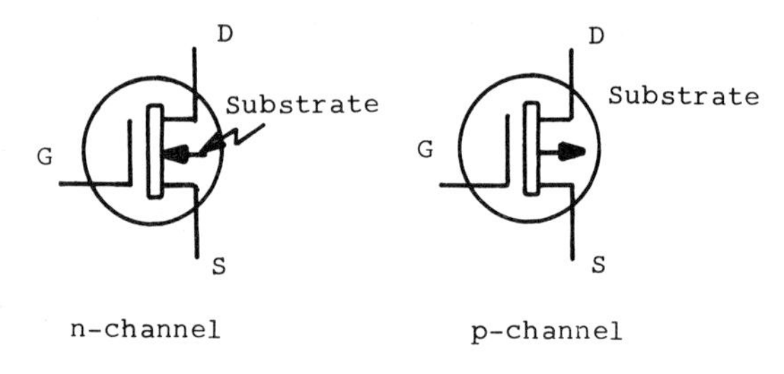

Fig. 4.11 Depletion-MOSFET symbols

The symbol, fig. 4.11, has a continuous line between drain and source, to indicate that there is normally a path between drain and source.

4.4 Comparison of FET's and bipolar transistors

There is considerable overlap in the suitability of JUGFET's, MOSFET's, and bipolar transistors for specific applications; for example, all three can be used as switches. However, there are distinct areas in which each type offers particular advantages. The most important differences stem from the different input impedances.

The choice of device for a particular application depends on the source resistance – the device resistance must 'match' the source to minimise drift (change in current or voltage due to temperature), noise (random unwanted voltage), and shunting (the short-circuiting effect of other circuits in parallel). If the source resistance is below a few thousand ohms, bipolar transistors are best; above this range to about 10 MΩ JUGFET's are preferred; above 10 MΩ MOSFET's are most suitable.

MOSFET's and JUGFET's make very good switches. They are not easily affected by noise, but are slower than bipolar switches. The small size and ease of construction combined with low power consumption make the MOSFET particularly suited for the manufacture of complex digital circuits, particularly memories and battery-powered devices such as watches.

The bipolar transistor has a much higher voltage gain, and the FET's cannot provide current or power gains. FET's are voltage-operated devices and, because of their larger inherent internal capacitances, FET's are not able to work at such high frequencies as the bipolar transistor. A reverse-biased p–n junction is an effective capacitor, thus the junction capacitance associated with the gate–source and gate–drain junctions of an FET is much greater than that for the equivalent terminals of a bipolar transistor.

4.5 Handling MOS devices

Since the insulation between the gate and channel of an MOS device is a very thin layer of silicon dioxide, effectively forming a capacitor, any charge applied between gate and source can cause extremely large electrical pressures across this effective dielectric. The 'static' which accumulates on the human body can be sufficient to 'punch through' the insulation if the leads are touched. Most devices do have some internal protection provided but it is not absolutely foolproof, so it is better to avoid touching the pins – in any case, the grease and oil from touching can cause poor connections to be made.

MOS devices are usually packed with their pins shorted by foil or conductive foam.

When soldering, use correct-temperature irons, preheated. In certain circumstances, specially earthed equipment and anti-static devices may be needed. Never solder or remove from a circuit with power on.

4.6 FET amplifiers

FET's can be used in many of the amplifier circuits normally using bipolar transistors. The selection of an operating point can be done using characteristics and load lines in an identical manner to that described for bipolar transistors in chapter 3.

4.7 Methods of bias

As FET's normally operate in class-A bias, some means of biasing is required. Batteries could be used but are inconvenient and expensive, so the following techniques are used.

4.7.1 Source self-bias

This technique is used with JUGFET's and depletion MOSFET's.

Figure 4.12(a) shows a typical common-source circuit. For a specified drain current, the required gate voltage V_{GS} can be found. Since gate current is negligible, the source resistor R_S can be found using Ohm's law:

$$R_S = V_{GS}/I_{DQ}$$

The resistor R_G provides a gate leakage path to prevent a build-up of charge on the gate when the signal is applied. The other components – R_D, C_{IN}, C_{OUT}, and C_S – carry out similar functions to those explained in chapter 3 for the common-emitter amplifier.

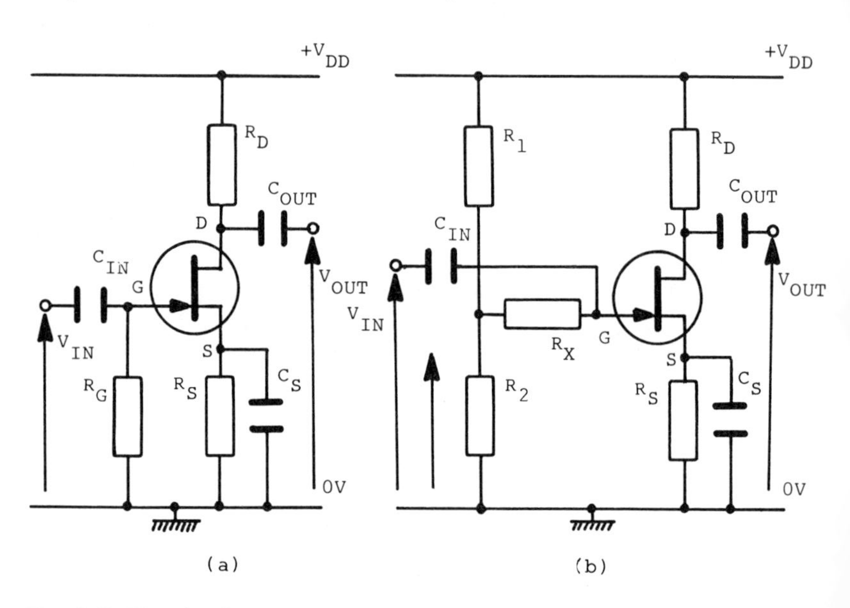

Fig. 4.12 Bias circuits: (a) source-self-bias circuit, (b) bias stabilisation with improved input impedance

The circuit has a number of disadvantages: bias takes a short time to build up while C_S charges up; the device cannot be biased to cut-off because some current must flow to keep C_S charged; and the bias point is subject to fluctuation with temperature.

4.7.2 Bias stabilisation

The circuit of fig. 4.12(b) provides stabilisation of the operating point over a wide temperature range. The bias voltage $V_{GS} = V_2 - I_D R_S$. The voltage V_2 can be assumed to be constant with changes in temperature. Changes in drain current due to temperature cause V_{GS} to increase or decrease by a suitable amount to correct the change, i.e. negative feedback (in a similar manner to that for the bipolar transistor in section 3.2.2).

For the assumption that V_2 is constant to be valid, the current through R_2 must be large compared to the gate current. In practice, this means that R_2 is quite a low-value resistor and, as it effectively shunts the FET, the circuit input resistance is effectively R_2. Since FET's are usually used because of their high input resistance, this can be a disadvantage. If the resistor R_X is included, the input resistance is now effectively R_X, but this can be omitted if a high input resistance is not important. Resistor R_X has no effect on the gate bias voltage, since the gate draws negligible current.

4.7.3 MOSFET bias

MOSFET's require the simplest arrangement for bias. Since the gate can be biased to the same potential as the drain, the gate can be connected to the supply rail.

4.8 Voltage gain of FET amplifier

The gain can be found using a load line on the characteristics in an identical way to that for bipolar transistors. Figure 4.13 shows a simple FET amplifier. The characteristics with the d.c. and a.c. load lines are shown in fig. 4.14.

The voltage gain of the amplifier, A_V, is given by

$$A_V = \frac{V_{DS2} - V_{DS1}}{V_{GS3} - V_{GS1}}$$

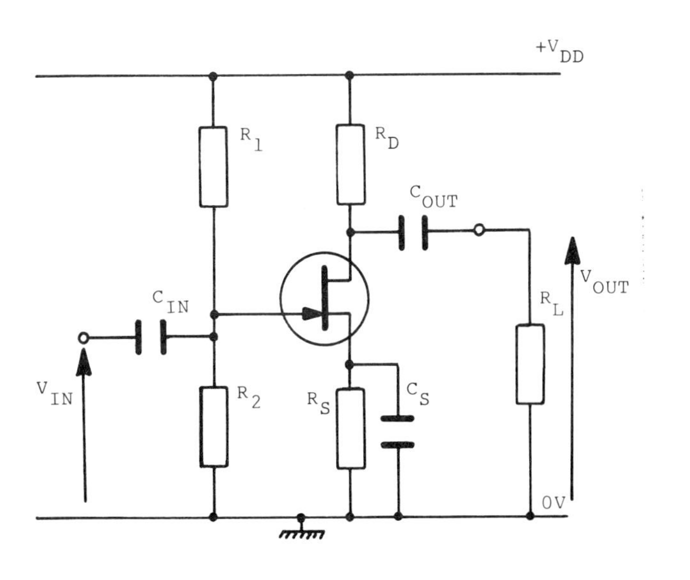

Fig. 4.13 Simple FET amplifier

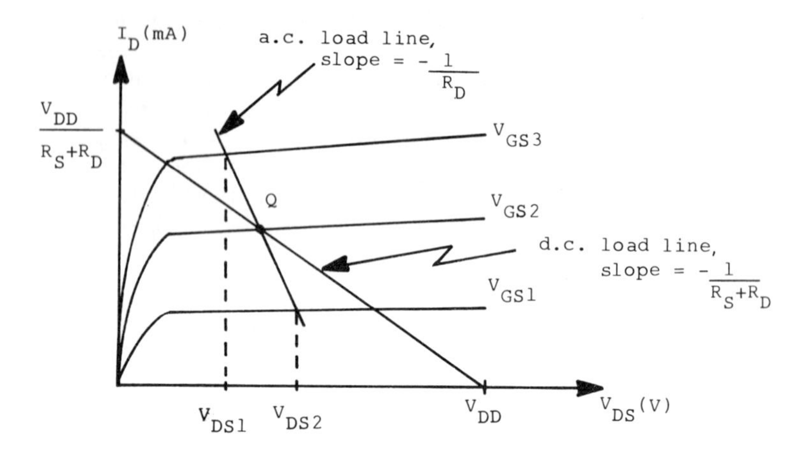

Fig. 4.14 Characteristics and load line for fig. 4.13

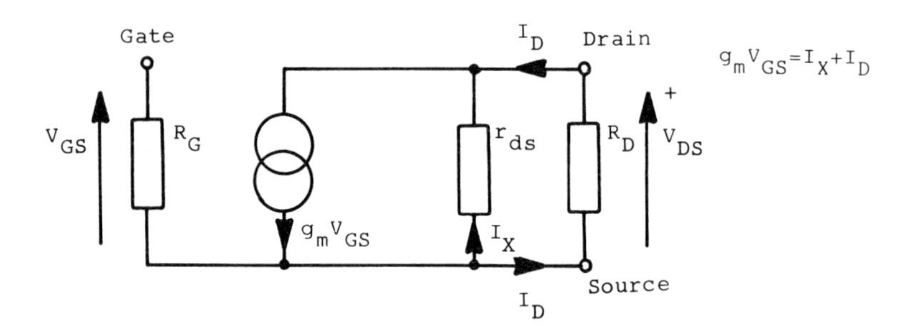

Fig. 4.15 Simple equivalent circuit for fig. 4.13

The FET can be represented by an 'equivalent circuit' (its mathematical model) to explain its small-signal operation – this is shown in fig. 4.15. The active part of the amplifier is represented by a 'constant-current' generator whose output is dependent on the input voltage V_{GS} and the mutual conductance g_m. Note that all capacitors and power supplies are effective short circuits over the operating range.

Now $g_m V_{GS} = I_X + I_D$

and $V_{DS} = -(I_X + I_D)R_P$

where $R_P = \dfrac{R_D \times r_{ds}}{R_D + r_{ds}}$

$$\therefore \quad V_{DS} = -g_m V_{GS} R_P$$

and gain $A_V = \dfrac{V_{DS}}{V_{GS}} = \dfrac{-g_m V_{GS} R_P}{V_{GS}} = -g_m R_P = \dfrac{-g_m R_D r_{ds}}{R_D + r_{ds}}$

If $r_{ds} > R_D$, which is normally the case, then A_V approximates to

$$A_V = \frac{-g_m R_D r_{ds}}{r_{ds}} \quad (\text{since } r_{ds} + R_D \approx r_{ds})$$

$$= -g_m R_D$$

Note: the minus sign in the above formula is required to justify the polarity chosen for V_{DS} in fig. 4.15 and it indicates that the output is phase-inverted.

Example 4.2 An FET has $g_m = 3$ mS and $r_{ds} = 40$ kΩ. If it is used with a drain load resistor R_D of 2.2 kΩ, calculate the voltage gain.

$$A_V = \frac{g_m r_{ds} R_D}{r_{ds} + R_D}$$

In this case, $r_{ds} > R_D$

$$\therefore \quad A_V \approx g_m R_D = 3 \text{ mS} \times 2.2 \text{ k}\Omega = 6.6$$

4.9 Bandwidth

The frequency-response curve for FET amplifiers is similar to that for bipolar transistors for the same reasons, except that the upper 3 db frequency is usually much lower, due to the FET having more internal capacitance.

4.10 Shunt frequency compensation of FET amplifiers

Some improvement in the frequency response can be achieved by the addition of an inductor in series with the load resistor R_D. The circuit arrangement is shown in fig. 4.16. If the value of the inductor is correctly chosen, the impedance of the load can be made to increase sufficiently over a range of frequencies and, since gain is normally dependent upon load impedance, some improvement in the upper cut-off frequency can be achieved to compensate for the fall in gain.

Example 4.3 Figure 4.17 gives the characteristics for a JUGFET to be used in the circuit of fig. 4.13. $V_{DD} = 12$ V, $R_D = 1$ kΩ, $R_S = 330$ Ω, and $R_L = 3$ kΩ.

a) Use the characteristics to find (i) the mutual conductance when $V_{DS} =$ 6 V, (ii) the drain–source resistance for $V_{GS} = -3$ V.
b) Use the values obtained in (a) to calculate the voltage gain for the circuit.
c) Draw the d.c. load line and select a suitable operating point.
d) Draw the a.c. load line and use it to find the voltage gain for an input signal of 1 V peak value.
e) Comment on any difference between the values obtained in (b) and (d) above.

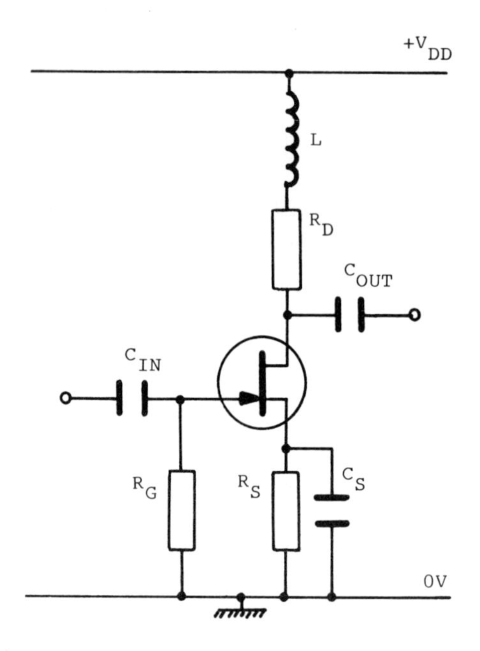

Fig. 4.16 Shunt frequency compensation

a) (i) $$g_m = \left.\frac{\Delta I_D}{\Delta V_{GS}}\right|_{V_D = 6\text{ V}} = \frac{y_2 - y_1}{4\text{ V} - 2\text{ V}}$$

$$= \frac{6.55\text{ mA} - 3.1\text{ mA}}{2\text{ V}} = 1.73\text{ mS}$$

ii) $$r_{ds} = \left.\frac{\Delta V_{DS}}{\Delta I_D}\right|_{V_{GS} = -3\text{ V}} = \frac{x_2 - x_1}{y_4 - y_3}$$

$$= \frac{9\text{ V} - 1\text{ V}}{5.6\text{ mA} - 3.6\text{ mA}} = 4\text{ k}\Omega$$

b) $$A_V = \frac{g_m r_{ds} R_P}{r_{ds} + R_P}$$

Note: R_D is in parallel with R_L:

$$R_P = \frac{R_D \times R_L}{R_D + R_L} = \frac{1\text{ k}\Omega \times 3\text{ k}\Omega}{4\text{ k}\Omega} = 750\ \Omega$$

$$\therefore\ A_V = \frac{1.73 \times 10^{-3} \times 4 \times 10^3 \times 750}{4 \times 10^3 + 750} = 1.1$$

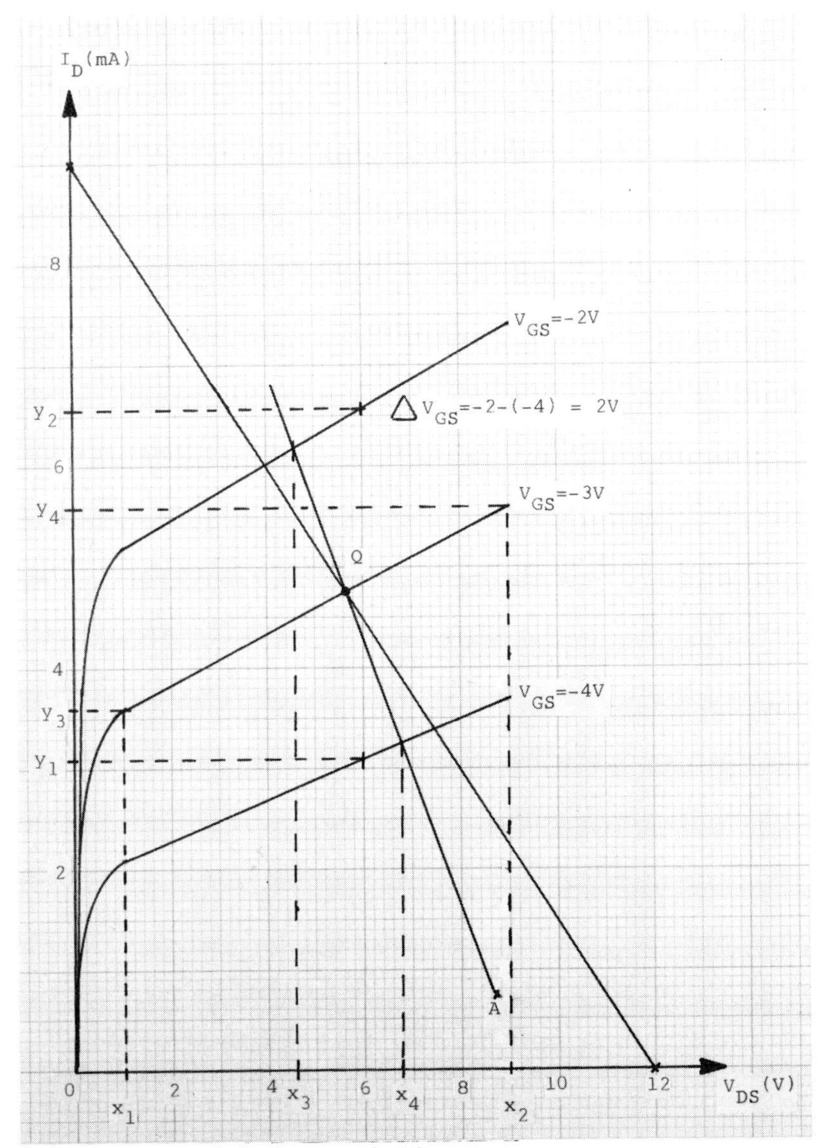

Fig. 4.17 Characteristics for example 4.3

c) The d.c. load line is positioned using

$$V_{DS} = V_{DD} = 12\text{ V} \quad \text{when} \quad I_D = 0$$

and $I_D = \dfrac{V_{DD}}{R_D + R_S} = \dfrac{12\text{ V}}{1.33\text{ k}\Omega} = 9\text{ mA} \quad \text{when} \quad V_{DS} = 0$

A suitable operating point is Q (5.7, 4.75).

d) The a.c. load line is positioned using 'Q' and $\Delta I_D/\Delta V_{DS} = -1/R_P$. Choose a convenient value for ΔV_{DS}, say 3 V; then

$$\Delta I_D = \frac{-3\text{ V}}{750\ \Omega} = -4\text{ mA}$$

(the minus sign indicates the direction of the slope).

The coordinates of the second point – 'A' on fig. 4.17 – are therefore given by $\{(5.7 + 3), (4.75 - 4)\} = (8.7, 0.75)$; thus the a.c. load line can be constructed.

For a 1 V peak input signal,

$$\Delta V_{GS} = 2\text{ V}$$

and $\Delta V_{DS} = x_4 - x_3 = 6.8\text{ V} - 4.6\text{ V} = 2.2\text{ V}$

$$\therefore \quad A_V = \frac{\Delta V_{DS}}{\Delta V_{GS}} = \frac{2.2\text{ V}}{2\text{ V}} = 1.05$$

e) The gain calculated in (d) above is approximately the same as that calculated in (b) – any differences can be due to errors in reading graphs and the fact that the value for g_m varies slightly for each value of V_{GS}.

Problems

The FET characteristics in figs 4.18, 4.19, and 4.20 may be photocopied for use by students or teachers.

4.1 A JUGFET has $g_m = 3.6$ mS and $r_{ds} \risingdotseq 100$ kΩ and is connected in a circuit with a drain load resistor of 4.7 kΩ. Calculate the voltage gain of the circuit.

4.2 A JUGFET has $g_m = 4$ mS and $r_{ds} = 25$ kΩ and is connected in a circuit with a drain load resistor of 5.6 kΩ. Calculate the voltage gain of the circuit.

4.3 The FET in fig. 4.13 has the drain characteristics given in fig. 4.18. $V_{DD} = 15$ V, $R_D = 1.1$ kΩ, $R_S = 300\ \Omega$, and the load R_L may be assumed to be infinite.

a) Draw the d.c. load line and choose a suitable bias point.

b) Draw the a.c. load line and, for an input signal of 0.5 V peak value, find (i) the voltage gain, (ii) the r.m.s. value of the a.c. component of the drain current.

4.4 The data in Table 4.1 was obtained for a BF245C JUGFET.

a) Plot the drain characteristics and use them to obtain (i) g_m for $V_{DS} = 12$ V, (ii) r_{ds} for $V_{GS} = -2$ V.

b) Using data obtained from the drain characteristics, plot the mutual characteristics for $V_{DS} = 14$ V and determine a value for g_m.

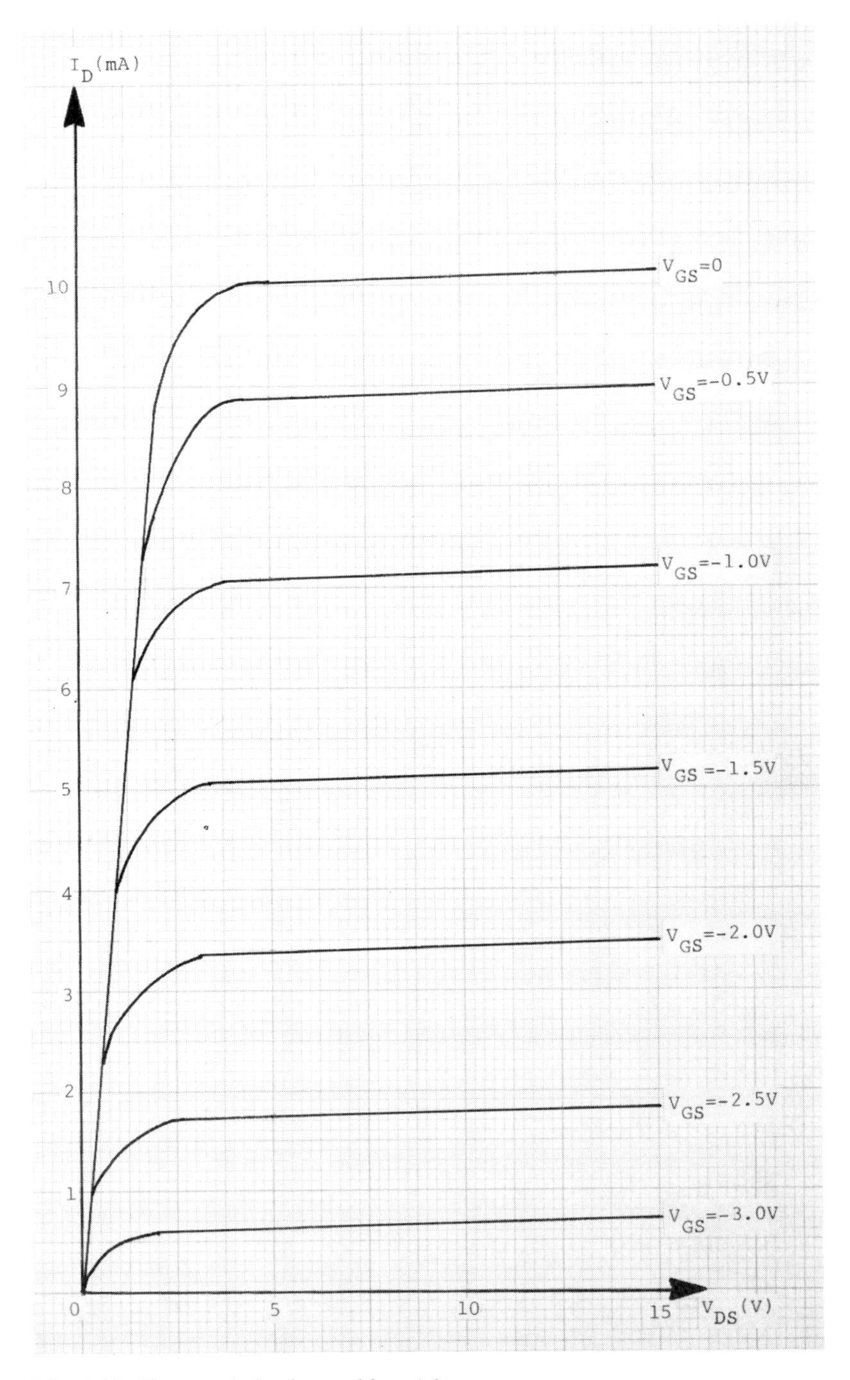

Fig. 4.18 Characteristics for problem 4.3

Table 4.1

	I_D (mA) for V_{GS} (V)		
V_{DS} (V)	0	−2	−4
5	16.25	8.0	2.8
10	16.8	8.25	2.9
15	17.5	8.5	3.0

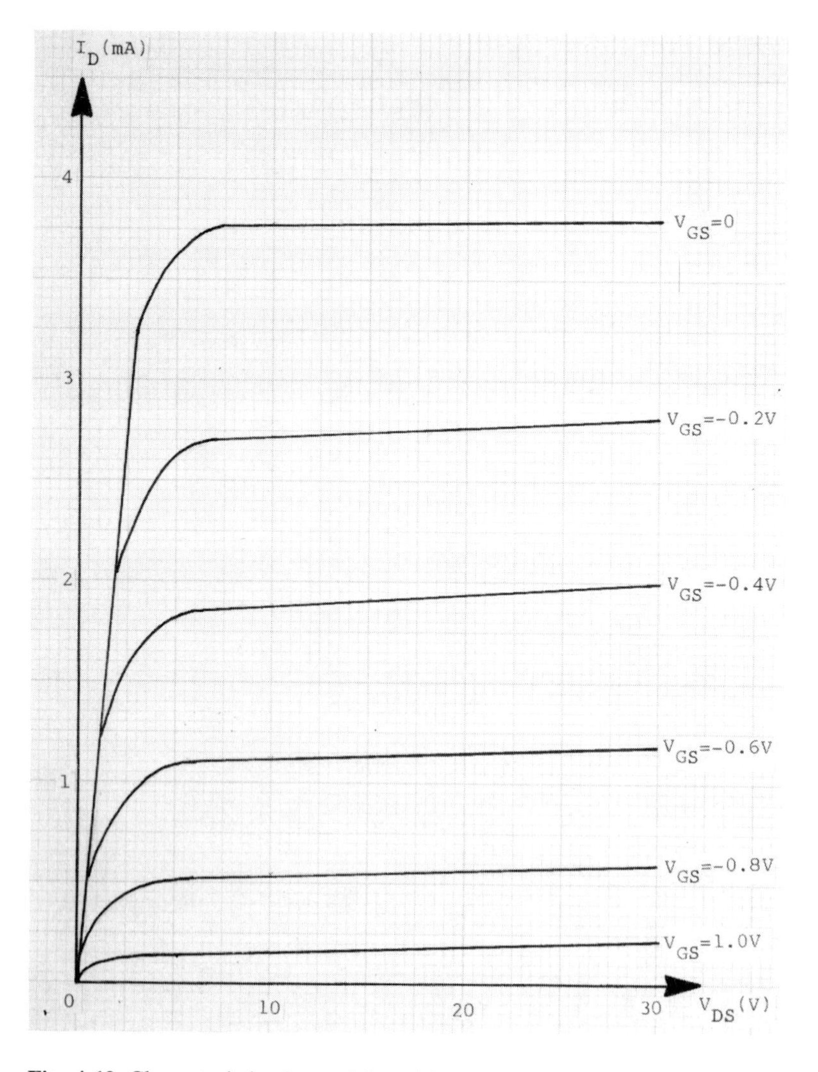

Fig. 4.19 Characteristics for problem 4.5

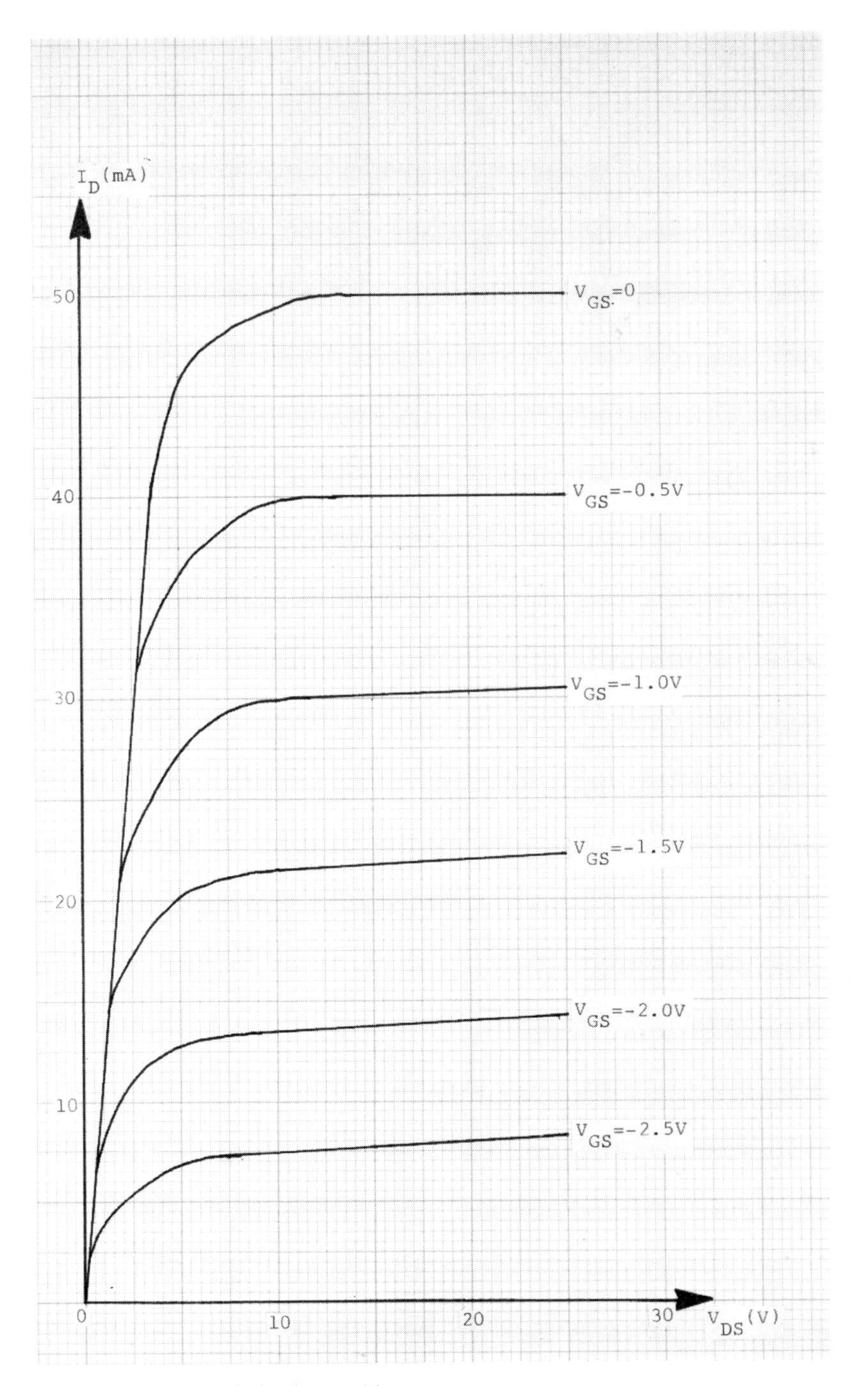

Fig. 4.20 Characteristics for problem 4.6

4.5 The JUGFET in fig. 4.13 has the characteristics in fig. 4.19. V_{DD} = 30 V, R_L = 3 kΩ, the quiescent drain current is 1.9 mA, and the quiescent drain voltage is 14 V.

a) Draw the d.c. load line and hence find the values of R_D and R_S. (Hint: find R_S first, using the bias voltage and quiescent drain current.)

b) Draw the a.c. load line and calculate the voltage gain for an input signal of 0.2 V peak value.

4.6 The circuit of fig. 4.13 uses a FET whose characteristics are given in fig. 4.20. V_{DD} = 24 V, R_1 = 560 kΩ, R_D = 300 Ω, R_S = 100 Ω, and R_L = 1.2 kΩ.

a) Draw the d.c. load line

b) If the gate bias voltage is to be – 1.5 V, draw the a.c. load line and calculate the voltage gain for an input signal of 0.25 V peak value.

c) Calculate the value of R_2. (Hint: the volt drop across R_2 can be found by considering the bias required and the actual p.d. across R_S with no signal.)

4.7 Explain the difference between 'pinch-off voltage' and 'pinch-off region' in a FET.

4.8 Explain the differences in construction between the JUGFET and the MOSFET.

4.9 Why is it important to ensure that the terminals of MOSFET's are protected by shorting them before use?

4.10 What class of bias would you normally expect an FET amplifier to operate in?

5 Power supplies

5.1 Introduction

Power supplies of any type, e.g. batteries or solar cells, are an essential part of any electronic system. There are four main types of power supply in use:

i) a.c. input, d.c. output;
ii) d.c. input, d.c. output;
iii) a.c. input, a.c. output;
iv) d.c. input, a.c. output.

Most industrial electronic equipment in use at present requires d.c., and type (i) is the most common power supply used, usually mains-rectified d.c.

All power supplies can be represented by the block-diagram layout of fig. 5.1.

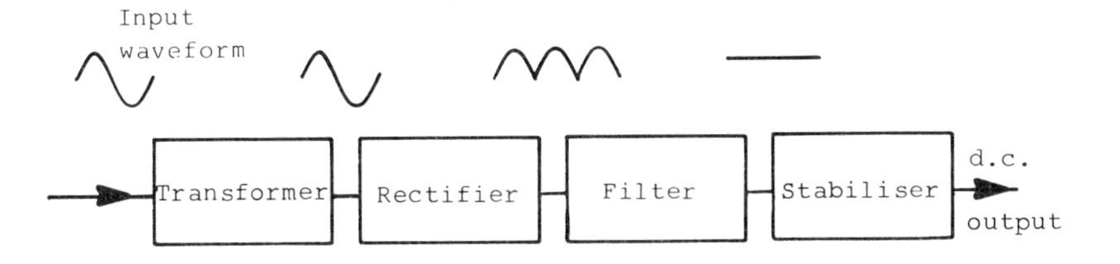

Fig. 5.1 Power-supply block diagram

5.2 Basic rectication

5.2.1 Half-wave rectification

The most simple form of unregulated power supply is the half-wave rectifier, shown with its input and output waveforms in fig. 5.2. Output current flows only during the positive half cycle, when the diode conducts. The output is a pulsating waveform of little practical use other than for battery charging.

5.2.2 Full-wave rectification

Full-wave rectifiers are more efficient and easier to smooth than half-wave rectifiers. There are two basic types: (a) bi-phase and (b) full-wave bridge.

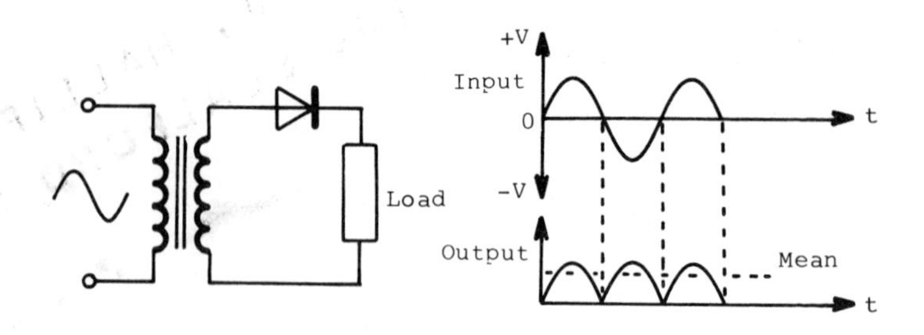

Fig. 5.2 Half-wave rectifier

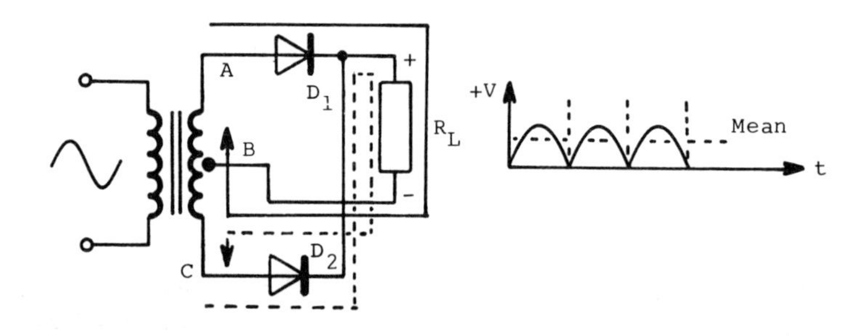

Fig. 5.3 Full-wave bi-phase rectifier

a) Bi-phase This circuit uses a centre-tapped transformer and two diodes and is shown with its output waveform in fig. 5.3.

If point A of the transformer secondary is positive with respect to B, then B is positive with respect to C. Diode D_1 will be forward-biased and current (shown by the continuous arrow) will flow through the load R_L from top to bottom and back to the transformer centre-tap; diode D_2 is reverse-biased. On the next half cycle, point C is positive with respect to B which is positive with respect to A; diode D_2 is now forward-biased and current (shown by a dotted arrow) will again flow through R_L from top to bottom back to point B; D_1 is reverse-biased.

b) Full-wave bridge The full-wave bridge produces an identical output to the bi-phase by using four diodes but no centre-tapped transformer. This makes it much cheaper than the bi-phase type and it will give twice the output for the same transformer. Figure 5.4 shows the basic circuit.

If point A on the transformer secondary is positive with respect to B, diodes D_2 and D_4 are forward-biased and current (shown by a continuous arrow) flows from D_2 to R_L top to bottom to D_4 to B. On the next half cycle, B is positive with respect to A and diodes D_1 and D_3 are forward-biased; current (shown by a dotted arrow) flows from D_3 to R_L top to bottom to D_1 to A. So the top of R_L is always positive. The diodes 'point'

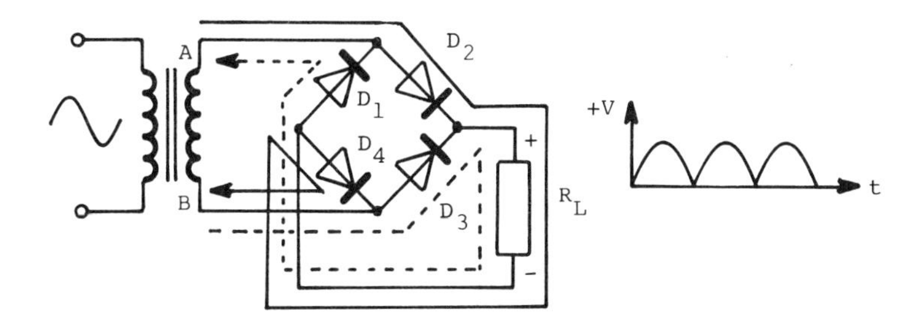

Fig. 5.4 Full-wave bridge rectifier

to the positive end of the load. The rectified output has a 'ripple' twice the frequency of that applied to the transformer, 100 Hz in the case of 50 Hz mains. The four diodes forming the bridge are available in a single composite package.

5.3 Smoothing or filter circuits

The output from the above full-wave rectifiers is not constant but is pulsating. This is of little practical use.

The simplest method of removing the worst of the 'lumps' is to use a large-value capacitor which acts as a 'reservoir' while the input wave is falling. The capacitor must have low reactance and, provided the time constant with the load is long compared to the supply period, a waveform similar to that in fig. 5.5 will result. Further smoothing can be achieved by adding resistors or inductors, but the advent of commercial regulating devices is making these types obsolete.

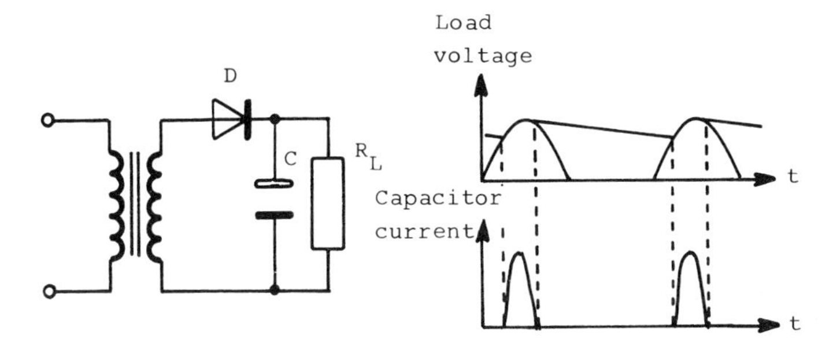

Fig. 5.5 Capacitor smoothing

5.4 Stabilised power supplies

Modern electronic circuits require power supplies which are much more stable than can be achieved by simple capacitor means. In any practical circuit, the output voltage may vary for a number of reasons; for example

a) fluctuations in mains supply;
b) variation of circuit parameters due to temperature, ageing, etc.;
c) variation in the load requirement.

5.4.1 Stabilising methods

a) Series stabilising The basic series-stabilising circuit is shown in fig. 5.6. The 'error-detector circuit' produces an error voltage, *e*, which is the difference between the output voltage V_{OUT} and a reference voltage V_{REF}. This error voltage is used directly to control the amount of current in the load R_L, and hence V_{OUT}.

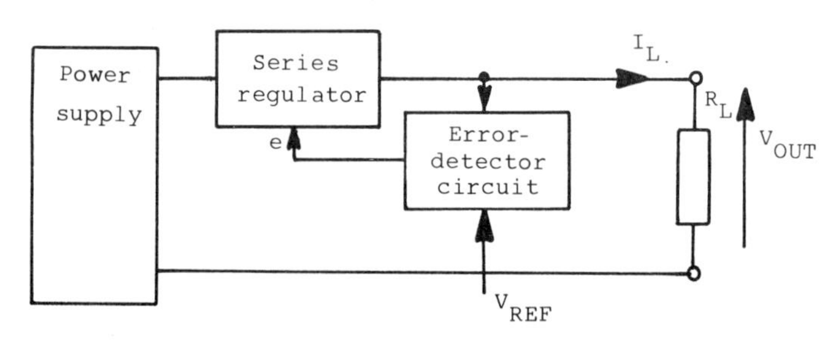

Fig. 5.6 Basic series stabiliser

Figure 5.7 shows a simple series-stabiliser circuit. Transistor T_1 is the series regulator and acts as variable resistor. If V_{OUT} decreases and the voltage at point A falls below the voltage of the reference zener plus the base–emitter voltage of T_2 (= 0.6 V), transistor T_2 will start to turn off. This will cause its collector potential – which is also the base potential of T_1 – to rise, turning T_1 on. As T_1 turns on, the volt drop across it reduces,

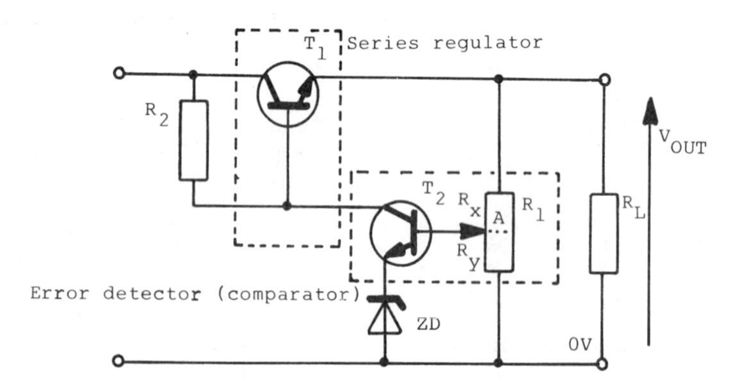

Fig. 5.7 Simple series-stabiliser circuit

increasing V_{OUT}. The opposite takes place if V_{OUT} rises above its desired level.

b) Parallel or shunt stabilising Figure 5.8 shows the basic arrangement for the shunt stabiliser. In this case, the error voltage is used to control a circuit to pass any excess current to ground, keeping the load current and hence $V_{OUT} = I_{OUT} \times R_L$ constant. The simplest regulator of this type is the zener stabiliser of fig. 1.13.

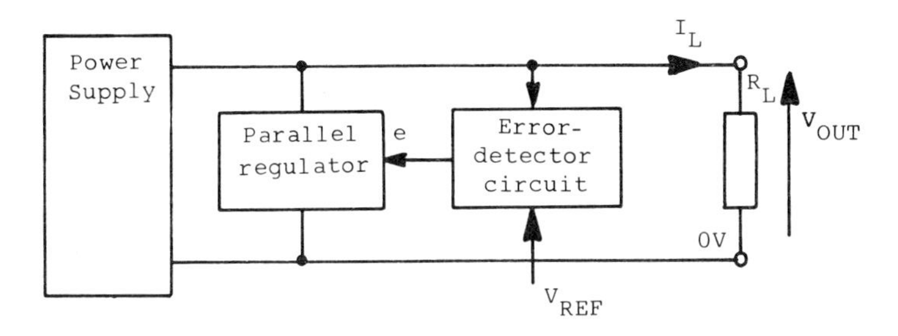

Fig. 5.8 Basic shunt stabiliser

In practice, the series stabiliser is preferred since it simply cuts down the current from the supply, whereas the shunt stabiliser bypasses any excess current to ground and the power supply *always* provides maximum output so the parallel regulator must be able to dissipate maximum power when $I_{OUT} = 0$.

5.5 Practical voltage regulators

These take a varying input voltage and produce a fixed regulated output. They are available in commercial packages with a choice of voltage and current ranges, with positive or negative outputs, e.g. an LM340T-5 is a + 5 V regulator in a TO220 package, rated at 1.5 A, input voltage range

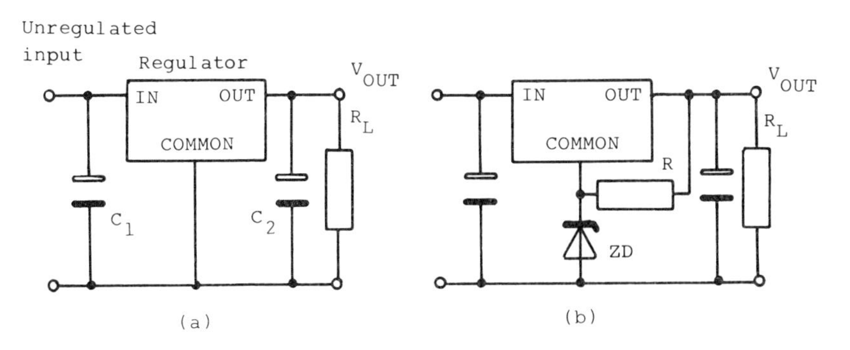

Fig. 5.9 Voltage regulator: (a) normal, (b) increased-output

7 V to 25 V. The basic circuit arrangement for regulators such as the LM340T-5 is shown in fig. 5.9(a). C_1 and C_2 are decoupling capacitors. C_1 prevents high-frequency instability; C_2, as well as decoupling the a.c. signal, reduces the output impedance at high frequencies.

The output of the basic regulator of fig. 5.9(a) can be increased by the addition of a zener diode as shown in fig. 5.9(b). Resistor R is a 'bleed' resistor to ensure that the zener diode is always operating in its active region.

5.6 Switched-mode power supplies

The basic regulators discussed previously are operated in the linear mode, i.e. somewhere between full on and full off. This means that the transistor itself is dissipating a large amount of power, and this dissipation is proportional to load current. Efficiencies greater than 50% are impossible.

Increasing use is now being made of switching regulators which can achieve efficiencies of 75%. They are particularly common in equipment such as microcomputers and cash registers which may be operating for long periods.

5.6.1 Basic switching regulator

Figure 5.10 shows the basic circuit of a switching regulator, in block-diagram form. The switch S is a solid-state device, such as a fast-switching transistor, which switches power on or off consuming little energy, depending on the error-detecting circuit. The inductor L – known as a 'swinging' inductor – stores energy while S is closed.

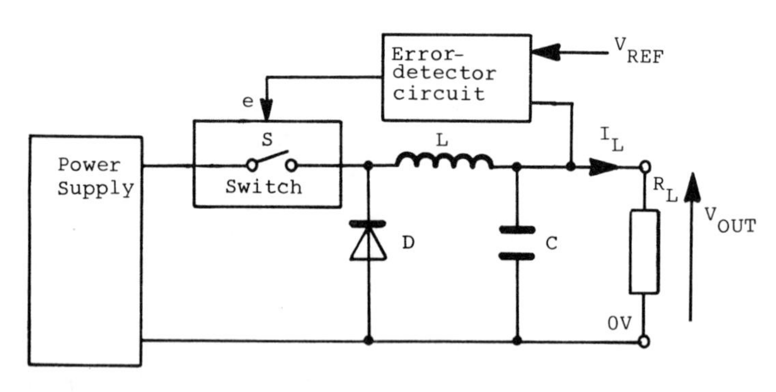

Fig. 5.10 Basic switching regulator

With S open, no current flows. With S closed, the full input voltage is applied across diode D (reverse-biased) and the full voltage appears across the inductor L and the current through it increases exponentially as capacitor C charges and raises the output voltage. When V_{OUT} reaches its predetermined value, the error-detector opens S and switches off the input current.

The inductor will prevent the current through it dropping to zero – since the magnetic energy stored must be dissipated, this will keep the current flowing through L in the same direction (Lenz's and Faraday's laws). Current now flows in the diode D, continuing to charge C and supply the load.

When the current supplied by L falls below the required value of I_{OUT}, the capacitor C supplies the load. This lowers the output voltage and eventually the error-detecting circuit switches S on again and the cycle is repeated.

Disadvantages of this circuit compared with those discussed in section 5.4 are higher 'ripple', poorer response to surges, noisier (hiss), and radio-frequency interference due to the switching action. Radio-frequency interference can be carried by the 'mains' or radiated (see chapter 6).

Problems

5.1 Explain why you would expect a mains a.c.-to-d.c. full-wave bridge rectifier to cost less than an equivalent full-wave bi-phase rectifier.

5.2 Give reasons why the output of a mains a.c.-to-d.c. full-wave rectifier will not be constant.

5.3 Why is a series-stabilised-regulator power supply preferred to a shunt (parallel) regulator?

6 Noise

6.1 Introduction

Strictly *noise* is defined as a random unwanted voltage, the name 'noise' coming from the fact that an amplifier with the input shorted and volume full on will produce an output 'hiss' or background noise. Modern usage has corrupted the definition to include any unwanted component in a signal, much of which is really 'interference'. We will adopt the newer definition of noise as a *random unwanted signal*. For practical purposes we will consider a 'signal' to consist of the wanted intelligence or data plus some unwanted noise.

Any signal passing through an amplifier system will normally be amplified – both the intelligence and the noise components – and some noise will be added by the amplifier itself. This is illustrated in fig. 6.1.

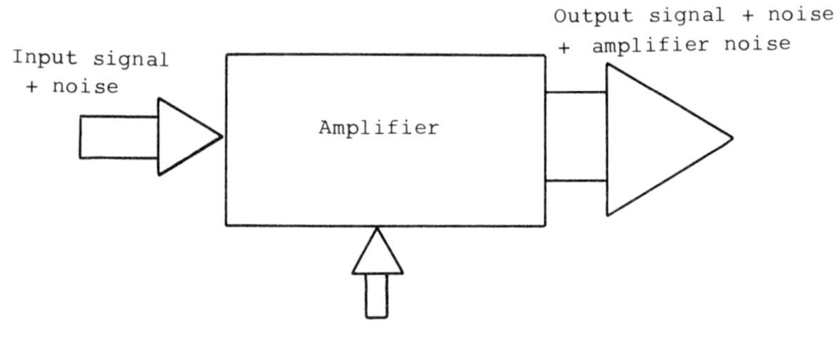

Fig. 6.1 Amplifier block diagram

While noise is usually an inconvenient by-product of normal operation, devices such as 'random number generators' can put the 'randomness' to good use; for example, 'Ernie' – the Post Office premium-bond machine – uses noise to select premium-bond winners.

6.2 Sources of noise

Because of our definition, noise can be conveniently be subdivided into two categories: (i) natural and (ii) man-made.

6.2.1 Natural noise

a) Thermal or white noise This is probably the most important type of noise, since it is produced by the random motion of electrons within every component. It is proportional to temperature and is truely random, i.e. the average noise power at any frequency is the same, thus noise is proportional to the amplifier bandwidth. Thermal noise provides a limit to the minimum signal which can be amplified, a major problem for radio astronomers trying to detect weak signals from space.

Thermal-noise voltage, E_N, can be calculated using

$$E_N = \sqrt{4RKTB} \quad 6.1$$

where E_N = r.m.s. noise voltage in volts
R = total circuit resistance in ohms
K = Boltzmann's constant = 1.38×10^{-23} joules/kelvin
T = absolute temperature in kelvin (°C + 273)
and B = bandwidth in hertz

Thermal noise can be reduced by operating the amplifier at low temperature and by using parametric amplifiers – these are special low-noise amplifiers usually using inductor–capacitor resonant circuits.

b) Solar radiation The sun, like all stars, is a radio emitter and is subject to cyclic variations in its radio output – the sun-spot cycle – which can play havoc with radio communications.

c) Galactic noise This is caused by radio emissions from stars in the galaxy and beyond, including the thermal background radiation due to the remnants of the 'big bang' (the gigantic explosion which is generally accepted to have occurred at the start of the universe).

d) Earth noise Thermal radiation from the earth produces noise which becomes more of a problem at frequencies above about 200 MHz. Atmospheric storms can affect communications and, via 'grid' power lines, can affect amplifiers which do not receive their input signal through an aerial.

e) Shot noise This is produced by the random division of current between base and collector in a transistor.

f) Barkhausen noise This is caused by the random movement of magnetic domains within a material.

g) Flicker noise This is proportional to $1/f$. No satisfactory explanation for it has yet been found.

h) Contact noise This is generated within certain types of resistive components, e.g. carbon, where electrons are moving from atom to atom.

Sources (b), (c), and (d) mainly affect communication receivers and can best be reduced by correct alignment of aerials, more directional aerials, and good receiver design. The other sources have to be lived with in all amplifiers, but their effects can be reduced by careful choice of components and operating environment, particularly temperature.

6.2.2 Man-made noise

a) Mains hum This is generated by mains-powered transformer rectifier systems.

b) Ignition or spark interference This is generated by any making or breaking electrical contacts, e.g. the v.h.f. radio interference generated by car ignition systems. Central-heating systems, refrigerators, and irons with thermostats can all produce spark interference, and it can either be radiated – emitted as radio waves, which does not require physical connection between the source and the effect – or be mains-borne – carried along the mains conductors and entering via the power supply.

c) Radio interference (r.f. interference) This is interference from an adjacent channel, due to overcrowding of the available radio spectrum.

d) X-ray interference This is radiated by all X-ray-producing equipment and mainly affects v.h.f. equipment.

e) Cross-talk This occurs when a signal on one set of cables is induced into another adjacent set. Modern digital equipment is particularly susceptible to this form of interference. It can be a difficult problem to solve where signal cables are laid close to mains cables, and it often occurs when rewiring or rerouting of mains cables takes place.

f) Device noise By the nature of their operation, particular devices such as thyristors (SCR's and triacs – see chapter 10) can generate mains-borne and radiated noise. In the case of thyristors, this can be almost totally eliminated by better firing techniques, e.g. zero-voltage switching (see section 10.3.3).

By far the best method of dealing with man-made noise external to the amplifier, is to eliminate it at source by using filters or 'suppressors' or, as in the case of car ignition systems, by changing 'noisy' components for less noisy ones (e.g. carbon plug leads rather than wire ones). Other techniques include electromagnetic and electrostatic shielding. Adjacent-channel

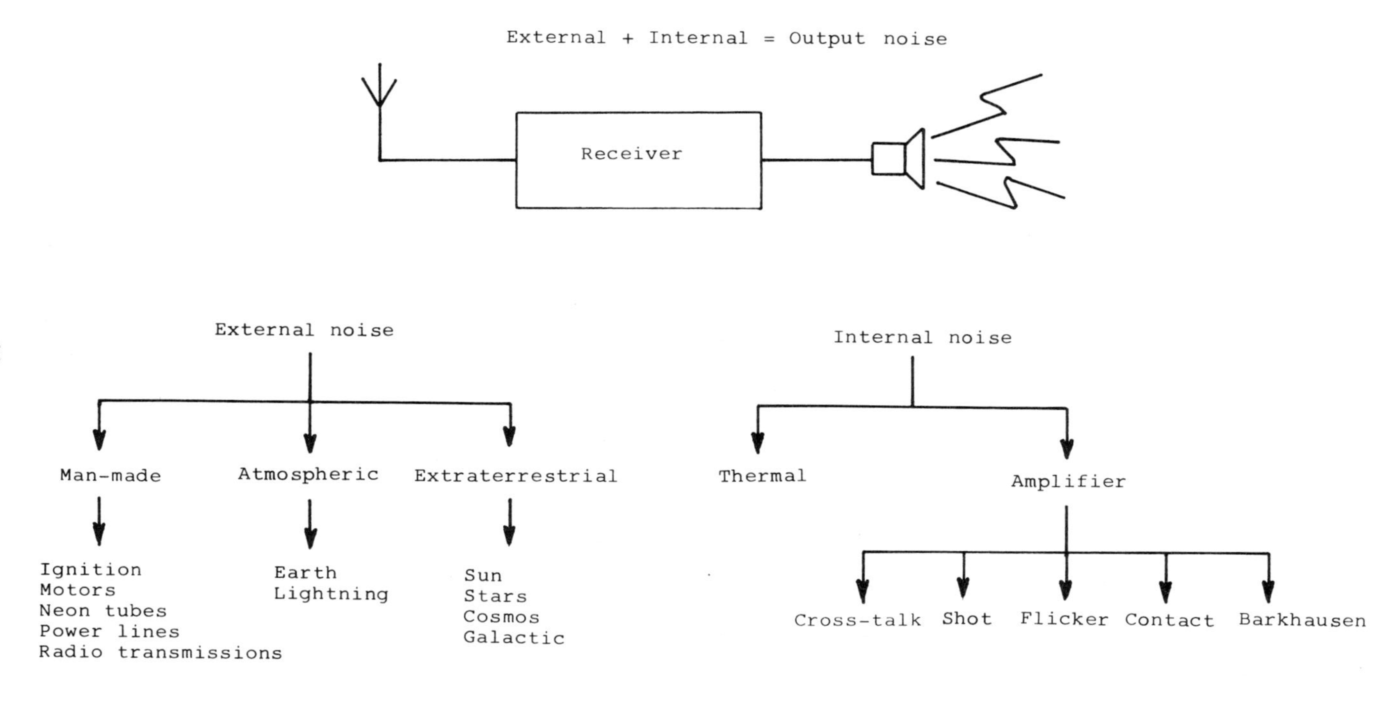

Fig. 6.2 Sources of noise

interference can be reduced by better receiver design and better channel allocation by the appropriate authorities. With many types of interference, such as cross-talk, careful attention to circuit design and layout are required.

The above lists are by no means exhaustive, particularly the man-made, but are simply intended to give the reader some idea of the problems. Figure 6.2 lists some of the above sources of noise.

6.3 Signal-to-noise ratio

Signal-to-noise ratio is the relationship between the wanted intelligence in a signal and the unwanted noise in the signal, and is defined as the ratio of the signal power (or voltage) to the noise power (or voltage) at a particular point in a circuit, for a given bandwidth. It is given by

$$\text{signal-to-noise ratio} = 10 \log_{10} \frac{\text{signal power}}{\text{noise power}} \text{ dB} \qquad 6.2$$

or, in terms of voltage,

$$\text{signal-to-noise ratio} = 20 \log_{10} \frac{\text{signal voltage}}{\text{noise voltage}} \text{ dB} \qquad 6.3$$

6.4 Noise factor (F)

As already stated, an amplifier will always add some noise when amplifying a signal. A 'figure of merit' used to indicate how good an amplifier is in this respect is the 'noise factor', F. This is sometimes referred to as the 'noise figure'. Noise factor is given by the ratio

$$F = \frac{\text{signal-to-noise ratio at the amplifier input}}{\text{signal-to-noise ratio at the amplifier output}} \qquad 6.4$$

$$= \frac{E_{SI}/E_{NI}}{E_{SO}/E_{NO}}$$

$$= \frac{E_{SI}E_{NO}}{E_{NI}E_{SO}}$$

or $$F = \frac{P_{SI}P_{NO}}{P_{NI}P_{SO}}$$

where E_{SI} and P_{NO} are the input signal voltage and the output noise power respectively, etc.

It is usual to quote the noise factor in decibels. F is then given by

$$F = 10 \log_{10} \frac{P_{SI}P_{NO}}{P_{NI}P_{SO}} \text{ dB} = 20 \log_{10} \frac{E_{SI}E_{NO}}{E_{NI}E_{SO}} \text{ dB}$$

The noise factor can be used to compare different amplifiers. Ideally the noise factor should be 0 dB. Practical amplifiers have a noise factor of typically 6 dB at low frequencies to 2 dB at 30 MHz.

It can be shown that the first stage in any amplifier determines the noise factor. As a result, a low-noise FET is often used as the input transistor.

Example 6.1 A circuit has a total resistance of 50 kΩ and a bandwidth of 20 kHz at an ambient temperature of 17°C. Determine the noise voltage generated.

Using equation 6.1,

$$E_N = \sqrt{4RKTB}$$
$$= \sqrt{(4 \times 50 \times 10^3 \times 1.38 \times 10^{-23} \times (273 + 17) \times 20 \times 10^3)} \text{ V}$$
$$= 4\ \mu\text{V}$$

Example 6.2 An oscilloscope having a bandwidth of 5 MHz is connected across the circuit in example 6.1. If the oscilloscope has a very high input impedance, determine the new noise generated.

The widest bandwidth is now 5 MHz and, since noise power is proportional to bandwidth, the noise voltage will increase.

$$E_N = \sqrt{4RKTB}$$
$$= \sqrt{(4 \times 50 \times 10^3 \times 1.38 \times 10^{-23} \times 290 \times 5 \times 10^6)} \text{ V}$$
$$= 63.3\ \mu\text{V}$$

Example 6.3 Determine the signal-to-noise ratio at the input of an amplifier if the signal voltage is 20 mV and the noise voltage is 5 μV.

Signal voltage E_S = 20 mV

Noise voltage E_N = 5 μV

$$\text{Signal-to-noise voltage ratio} = 20 \log_{10}(E_S/E_N) \text{ dB}$$
$$= 20 \log_{10}(20 \times 10^{-3}/5 \times 10^{-6})$$
$$= 20 \log_{10}(4 \times 10^3)$$
$$= 72.2 \text{ dB}$$

Example 6.4 The amplifier in example 6.3 adds an extra 5 μV of noise to the output signal. If the gain of the amplifier is 12 dB, determine (a) the output signal-to-noise ratio in dB, (b) the noise factor of the amplifier in dB.

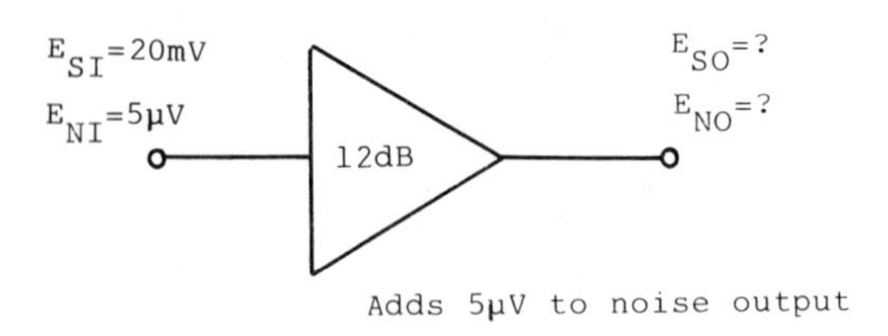

Fig. 6.3 Diagram for example 6.4

a) The system is shown in fig. 6.3.

$$\text{Gain in dB} = 12 = 20 \log_{10} (V_{OUT}/V_{IN})$$
$$= 20 \log_{10} A_V$$
$$\therefore \quad 12/20 = \log_{10} A_V$$
$$\therefore \quad A_V = \text{antilog } (12/20)$$

Solving this equation for A_V gives $A_V = 4$

$$\therefore \quad E_{SO} = E_{SI} \times A_V = 20 \text{ mV} \times 4 = 80 \text{ mV}$$

and $E_{NO} = E_{NI} \times A_V = 5 \ \mu\text{V} \times 4 = 20 \ \mu\text{V}$

$\therefore$ total noise output voltage $= 20 \ \mu\text{V} + 5 \ \mu\text{V} = 25 \ \mu\text{V}$

$\therefore$ output signal-to-noise ratio $= 20 \log_{10} (80 \times 10^{-3}/25 \times 10^{-6})$

$$= 70.5 \text{ dB}$$

The signal-to-noise ratio has fallen due to the amplifier noise.

b) $$F = 20 \log_{10} \frac{E_{SI}E_{NO}}{E_{NI}E_{SO}} \text{ dB} = 20 \log_{10} \frac{20 \times 10^{-3} \times 25 \times 10^{-6}}{5 \times 10^{-6} \times 80 \times 10^{-3}} \text{ dB}$$
$$= 20 \log_{10} 1.25 \text{ dB}$$
$$= 1.9 \text{ dB}$$

Problems

6.1 Calculate the signal-to-noise ratio for the signal and noise values given in Table 6.1.

6.2 An amplifier has a signal-to-noise ratio of 30 dB and a noise power of 25 μW. Calculate the signal output power.

6.3 The input to an amplifier consists of a wanted-signal component of 22 mW and a noise-power component of 2 μW. The power gain of the amplifier over the range of operation is constant at 25. The amplifier generates internal noise of 200 μW measured at the output terminals. Calculate the signal-to-noise ratio at the input and output of the amplifier.

Table 6.1

Signal	Noise	S/N ratio
1 mW	1 μW	
25 mW	1 μW	
4 mW	4 mW	
10 mW	20 mW	
8 mV	12 μV	
1 V	1 mV	

6.4 Table 6.2 gives the input and output voltages of an amplifier system at various frequencies.

a) Using a logarithmic scale for frequency, plot the gain/frequency characteristic.
b) Determine the amplifier bandwidth.
c) Find the noise voltage generated if the amplifier impedance can be considered to be 2 kΩ.

Table 6.2

Frequency (Hz)	10	100	1 k	10 k	100 k	1 M
Input voltage (mV)	20	20	20	20	20	20
Output voltage (mV)	800	1600	2000	2000	1700	1000

6.5 Define the term 'noise' in relation to electrical circuits.
6.6 Explain why it is impossible to eliminate all sources of noise in an amplifier.
6.7 What is the reason for using a 'signal-to-noise' ratio?
6.8 Explain why the signal-to-noise ratio at the output of an amplifier is less than the signal-to-noise ratio at the input.

7 Feedback

7.1 Introduction

If a small part of the output signal of an amplifier is fed back via an external circuit and is connected in series with the input signal, the operation of the circuit can be drastically altered.

Feedback can be either *negative* (degenerative) when the fed-back signal is in anti-phase to the input signal or *positive* (regenerative) when the fed-back signal is in phase with the input signal. Positive feedback has its main application in oscillators.

It should be noted that feedback may occur inadvertently with most amplifiers, and uncontrollable oscillations can occur if the conditions are right. This should not be confused with feedback deliberately introduced to achieve some specific end.

The amount of signal fed back can be made proportional to either the output current or the output voltage and can be fed back in series or parallel with the input signal.

7.2 Negative feedback

This is used in virtually all modern amplifiers. Figure 7.1(a) shows the block diagram of an amplifier in which a fraction β of the output is fed back via a network. The feedback network is usually a resistive network, but at this stage it is not important.

Definition of terms used

A = amplifier gain without feedback = open-loop gain

β = fraction of the output signal to be fed back
= feedback fraction

V_S = signal voltage to the circuit

V_{OUT} = amplifier output voltage

βV_{OUT} = the fed-back voltage

V_{IN} = input signal to the amplifier = $V_S + \beta V_{OUT}$

$A\beta$ = the loop gain

$1 + A\beta$ = feedback factor

For convenience, the circuit of fig. 7.1(a) is often replaced by a single-line diagram as shown in fig. 7.1(b).

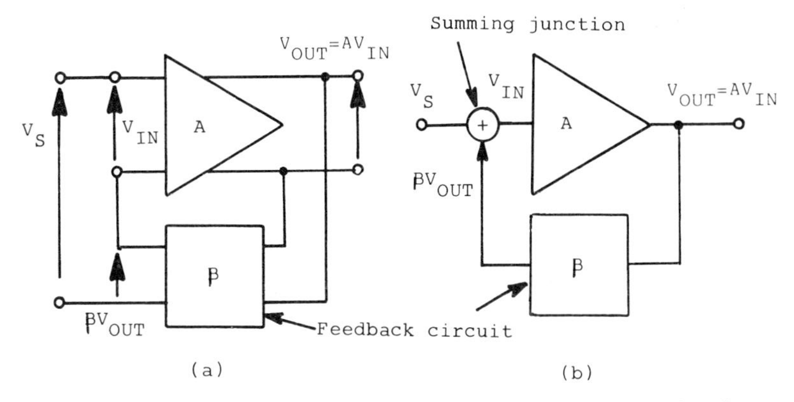

Fig. 7.1 Basic feedback arrangement: (a) block diagram, (b) single-line diagram

We can now derive the *general* feedback equation for the circuit.

$$V_{IN} = V_S + \beta V_{OUT}$$

and $$V_{OUT} = AV_{IN}$$

$$\therefore \quad V_{OUT} = A(V_S + \beta V_{OUT}) = AV_S + A\beta V_{OUT}$$

Transposing gives

$$V_{OUT} - A\beta V_{OUT} = AV_S$$

$$\therefore \quad V_{OUT}(1 - A\beta) = AV_S$$

Dividing both sides by $V_S(1 - A\beta)$ gives

$$\frac{V_{OUT}}{V_S} = \frac{A}{1 - A\beta}$$

Thus the overall gain of the amplifier *with* feedback, A', is given by

$$A' = \frac{A}{1 - A\beta} \qquad 7.1$$

This is the *general* feedback equation.

If the feedback is negative (i.e. in opposition to the input signal), then β is negative and equation 7.1 becomes

$$A' = \frac{A}{1 - A(-\beta)} = \frac{A}{1 + A\beta} \qquad 7.2$$

By inspection of equation 7.2 we can see that the gain with negative feedback, A', must always be less than the gain without feedback, A, since the denominator is greater than 1; i.e. negative feedback reduces overall gain. This might seem a bit self-defeating in an 'amplifier', but this is far outweighed by the benefits obtained (see below).

If the gain of the amplifier is very large (as it is for an operational amplifier) then the denominator of equation 7.2 effectively becomes $A\beta$ (since $A\beta \gg 1$). Then

$$A' \approx \frac{A}{A\beta} = \frac{1}{\beta} \qquad 7.3$$

i.e., provided A is very large, the gain of the system is *independent* of the gain of the amplifier and is dependent only on β. β can be made very stable using simple resistive components.

7.3 Effect of negative feedback on gain

The effect of negative feedback on gain is most easily shown by the use of an example. An amplifier with an open-loop gain of 200 000 (a 741 operational amplifier – see chapter 11) is used with 1% (not a large amount) negative feedback. The gain A' is

$$A' = \frac{A}{1 + A\beta} = \frac{200\,000}{1 + 200\,000 \times 0.01} = 99.95 \approx 100$$

If, due to ageing, the gain of the amplifier falls to 100 000, then

$$A' = \frac{100\,000}{1 + 100\,000 \times 0.01} = 99.90 \approx 100$$

Hence a large reduction in gain A has little or no effect on A'. For semiconductor devices which are subject to a large 'spread' in their parameters (for a BC109, h_{fe} = 200 to 800) and to variations in gain due to temperature, this is extremely useful. Negative feedback reduces and stabilises amplifier gain.

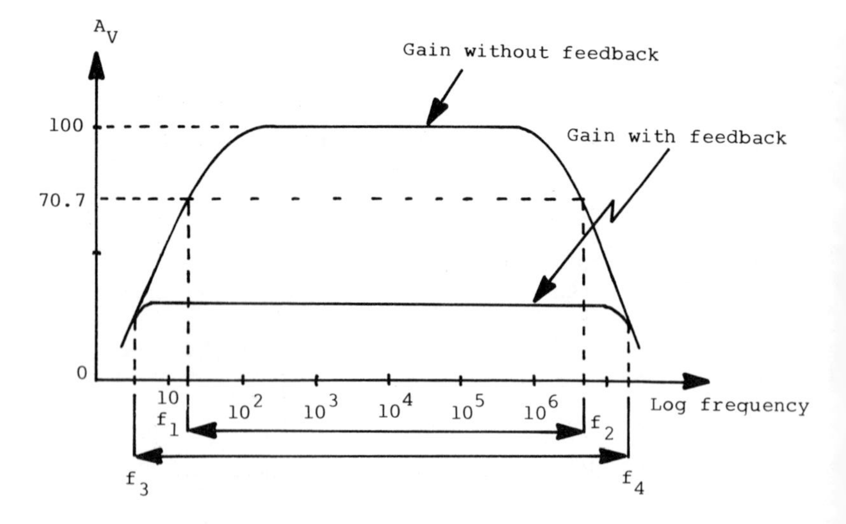

Fig. 7.2 Gain/frequency characteristic with and without feedback

7.4 Effect of negative feedback on bandwidth

Figure 7.2 shows the gain/frequency characteristic of a typical amplifier with and without feedback. The bandwidth without feedback is $f_2 - f_1 =$ 5 MHz − 25 Hz ≈ 5 MHz and with feedback is $f_4 - f_3 =$ 25 MHz − 3 Hz ≈ 25 MHz. Negative feedback increases bandwidth.

7.5 Effect of negative feedback on distortion and noise

Distortion and noise produced *within* the amplifier are reduced by negative feedback. Noise in any signal when it arrives at the amplifier *will not* be reduced.

7.6 Effect of negative feedback on input and output impedances

Although the derivations are beyond the scope of this book, it can be shown that, depending on which of the four types of feedback – e.g. voltage series, current shunt, etc. – is used, the input and output impedances can be modified to almost any desired value.

7.7 Practical circuits

In chapter 3, the operation of a common-emitter amplifier with base-voltage bias was explained. Any change in I_B results in a change in I_E which tries to prevent the change by altering the base–emitter voltage. This is an example of negative feedback using 'current-series feedback' ($\beta = R_E/R_C$).

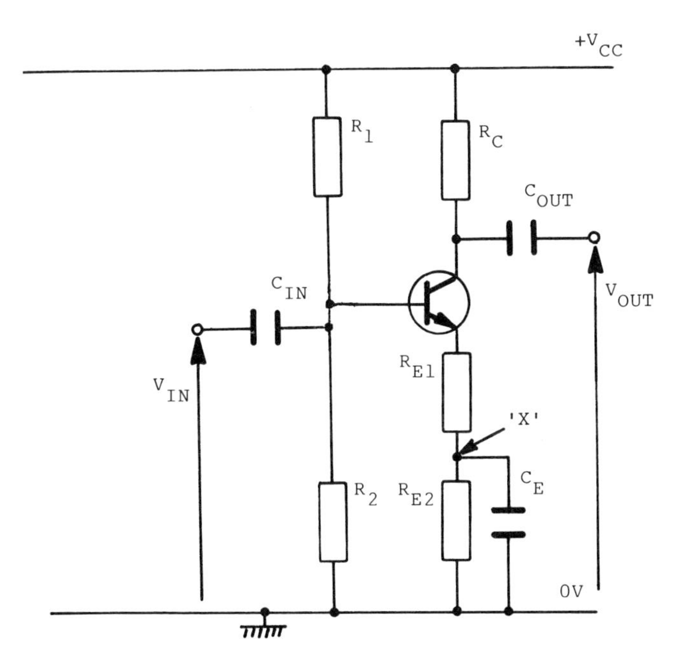

Fig. 7.3 Unbypassed emitter resistor

If negative feedback is required to affect the a.c. signal, an unbypassed resistor could be added as shown in fig. 7.3. In this case, for d.c. the emitter resistor is effectively $R_{E1} + R_{E2}$; for a.c., point 'X' is grounded.

7.8 The emitter follower

Another application using negative feedback is the emitter follower (also known as the common-collector or grounded-collector amplifier). The basic circuit is shown in fig. 7.4(a). The output is taken from the emitter. The input, as in the common-emitter amplifier, is V_{BE}. All of the output voltage is fed back, giving a gain of approximately unity. The input impedance is very high and can be shown to be approximately $h_{fe} R_E$. The FET equivalent of the emitter follower – the source follower – is shown in fig. 7.4(b).

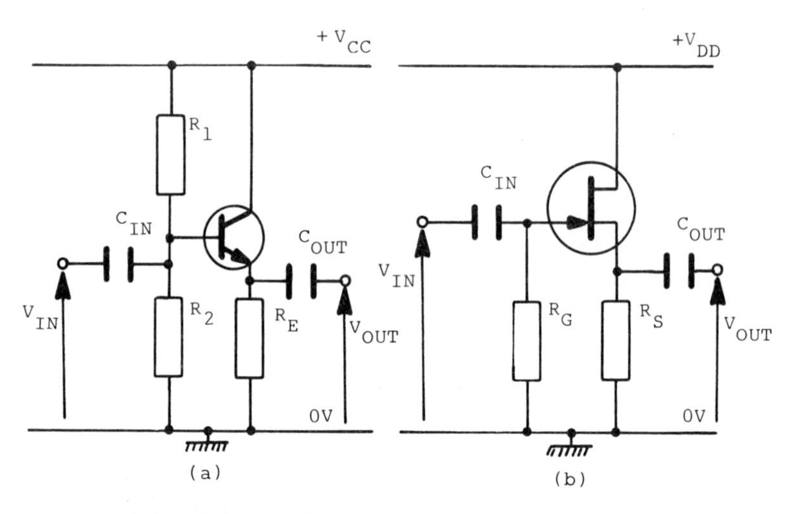

Fig. 7.4 (a) Emitter follower, (b) source follower

Example 7.1 The inverting amplifier in fig. 7.5 has a gain without feedback of 500. Part of the output signal is fed back via a resistive feedback network, the potential-divider chain R_1 and R_2. If $A = 500$, $R_1 = 270$ kΩ, and $R_2 = 3.3$ kΩ, calculate

a) the feedback fraction β,
b) the gain of the system with feedback.

a) Feedback fraction $\beta = \dfrac{R_2}{R_1 + R_2} = \dfrac{3.3 \text{ k}\Omega}{270 \text{ k}\Omega + 3.3 \text{ k}\Omega} = 0.012$

b) Gain $A' = \dfrac{A}{1 + \beta A} = \dfrac{500}{1 + 0.012 \times 500} = 71.43$

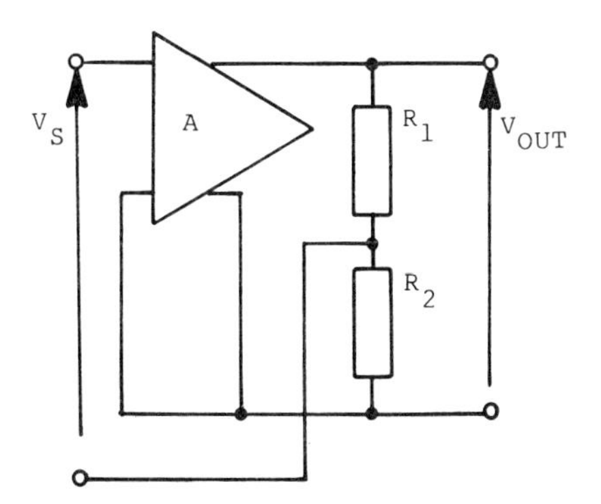

Fig. 7.5 Circuit for example 7.1 and problem 7.3.

Example 7.2 An inverting amplifier has a gain without feedback of 1000. Calculate the feedback fraction β required to give a stabilised gain of 25.

$$A' = \frac{A}{1 + \beta A}$$

Transposing gives

$$\beta = \frac{A - A'}{AA'} = \frac{1000 - 25}{1000 \times 25} = 0.039$$

Problems

7.1 An amplifier has a gain with negative feedback of 25. The gain without feedback is 10 000. Calculate the feedback fraction β.

7.2 An amplifier has a gain without feedback of 20 000. If 2.5% of the output is fed back in anti-phase to the input, calculate the resultant gain.

7.3 The amplifier of fig. 7.5 has a gain $A = 20\ 000$ and is required to produce a gain with negative feedback of 100.

a) Calculate the value of R_2 which will achieve this if $R_1 = 82\ \mathrm{k}\Omega$.

b) Using the nearest preferred value in the 5% tolerance range of resistors for R_2, calculate the gain of the system if the gain without feedback falls to 10 000.

7.4 The data in Table 7.1 applies to an amplifier without feedback.

a) Plot the gain/frequency characteristic and determine the amplifier bandwidth.

b) If negative feedback with a feedback fraction of 0.05 is applied, plot the new gain/frequency characteristic and determine the new bandwidth.

Table 7.1

Frequency (Hz)	10	100	1 k	10 k	100 k	1 M
Gain	25	75	100	100	80	40

7.5 An inverting amplifier has a voltage gain of 5000 without feedback. Negative feedback with a feedback fraction of 0.02 is used. Find

a) the feedback factor,
b) the loop gain,
c) the gain with feedback,
d) the percentage fall in gain with feedback if the amplifier gain without feedback falls by 25%.

7.6 Explain what advantages, offsetting the reduction in gain, are achieved by using negative feedback in an amplifier.

7.7 Derive the equation for gain for a circuit using negative feedback.

8 Direct-coupled amplifiers

8.1 Introduction

There are applications, particularly in control, when it is necessary to amplify d.c. or slowly varying a.c. signals. As explained in chapter 3, the reactance of any capacitors used for coupling causes the gain to fall off at low frequencies, so *R*–*C*-coupled amplifiers cannot be used. Special coupling techniques are required.

The most common method used is 'direct coupling', where the output of one stage is connected directly to the next stage. A less common method is to 'chop up' the d.c. into a square wave, amplify it in a conventional a.c. amplifier, and then rectify it.

In a.c. amplifiers, the smallest signal which can be amplified is determined by the noise generated in the first stage. In direct-coupled, or just 'D.C.', amplifiers, a limit is reached before this due to *drift*.

Drift can be defined as the change in performance of an amplifier (or transistor) due to changes in current gain, leakage current, or V_{BE} due to changes in temperature, changes in supply voltage, or ageing. The major factor is change in V_{BE} due to temperature – for this reason, silicon devices are preferred because of their lower initial leakage current.

Various techniques are used to try to minimise drift – feedback may be used, but this will also affect the signal; compensating signals can generated using heat-sensitive elements; or a design using stable low-noise input stages can be adopted.

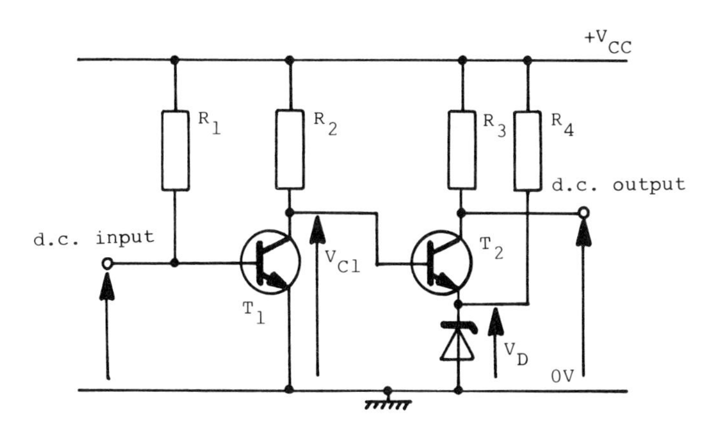

Fig. 8.1 Simple direct-coupled amplifier

The simplest circuit is shown in fig. 8.1. It consists of two common-emitter amplifiers. Resistor R_1 biases the first transistor T_1. Resistor R_2 provides the load for T_1 and also bias for transistor T_2. Resistor R_3 is the load for T_2. The zener diode is used in the emitter lead of T_2 to raise the base voltage of T_2 and hence the collector of T_1 to a reasonable level. (A resistor in place of the zener diode would have the same effect on the potentials but would introduce negative feedback when the current changed. A zener diode can maintain the voltage across it constant over a large range of current.) The resistor R_4 ensures that the zener diode is always biased into its active region even if the input to T_2 is zero. The zener voltage is chosen to be approximately 0.6 V below the required collector potential for T_1.

8.2 The Darlington pair

A very useful combination forming a D.C. amplifier is the 'Darlington pair'. This combines a very high input impedance with a low output impedance and high current gain. Figure 8.2 shows one arrangement using two discrete transistors, although this arrangement is available as a composite device. The arrangement consists of an emitter follower T_1, which gives it its high input impedance, and a common-emitter amplifier T_2, which determines the voltage gain.

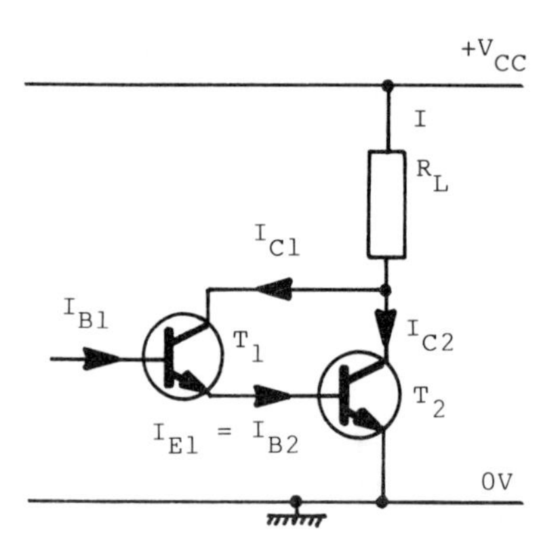

Fig. 8.2 Darlington-pair arrangement

The overall current gain, I/I_{B1}, can be found as shown below:

$$I_{E1} = I_{B1} + h_{FE1}I_{B1} \approx h_{FE1}I_{B1}$$

The emitter current of T_1 is the base current of T_2,

$$\therefore \quad I_{C2} = h_{FE2}I_{B2} = h_{FE2}h_{FE1}I_{B1}$$

Now $I = I_{C1} + I_{C2}$

but, if h_{FE2} is large, $I_{C2} \gg I_{C1}$

$$\therefore \quad I \approx I_{C2}$$

$$\therefore \quad \frac{I}{I_{B1}} = \frac{I_{C2}}{I_{B1}} = \frac{h_{FE2}h_{FE1}I_{B1}}{I_{B1}}$$

$$= h_{FE1}h_{FE2} \qquad 8.1$$

Due to its high current gain, this circuit is subject to drift as the leakage current of T_1 is amplified by T_2.

The circuit can be modified by putting the load resistor R_L in the emitter lead of T_2, putting both transistors in the emitter-follower mode. This arrangement gives an extremely high input impedance.

8.3 Differential amplifier

A circuit which has little drift is shown in fig. 8.3 – the 'long-tailed pair' or 'emitter-coupled pair'. The transistors should be 'matched', i.e. chosen so that they are as far as is possible identical in all respects, particularly to see that they have equal changes of V_{BE} with temperature change. Matched pairs are commercially available. The other circuit components should be symmetrical; i.e. $R_1 = R_3$, $R_2 = R_4$, and both collector load resistors R_C should be identical.

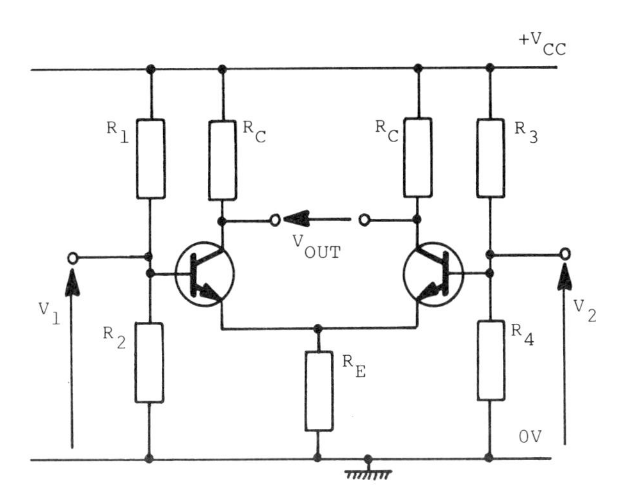

Fig. 8.3 Emitter-coupled pair

Biasing is achieved by the potential-divider networks R_1, R_2 and R_3, R_4. Resistor R_E is a large-value resistor to fix V_E, and $I_{RE} \approx I_{C1} + I_{C2}$.

If identical voltages, in both amplitude *and* phase, are applied to both inputs, then both collector currents and hence both collector voltages will change; but if the transistors are identical then V_{OUT}, which is the difference between the two collector potentials, will be zero. Identical input signals of this form are called 'common-mode'.

If the input voltages are not the same, the collector currents will be different and V_{OUT} will be the difference between the two collector voltages and will be an amplified version of the input voltage difference.

Any changes in V_{BE} and I_{CBO} due to temperature, ageing, supply-voltage variations, etc. will cause common-mode changes and hence no output. It is not necessary to decouple R_E for a.c. since I_{RE} is constant for any difference in inputs – the collector currents increase and decrease by equal amounts. Ideally, both transistors should be mounted close together on the same heat sink. In the event of a failure, *both* transistors must be changed.

In practice, differential amplifiers are often made using integrated circuits.

Problems

8.1 Explain why a conventional *R–C*-coupled amplifier cannot be used for amplifying d.c. or slowly varying a.c. signals.

8.2 Define the term 'drift' in relation to direct-coupled amplifiers.

8.3 What are the advantages of using a Darlington-pair arrangement as a direct-coupled amplifier?

8.4 What is a 'common-mode' signal?

9 Power amplifiers and tuned amplifiers

9.1 Introduction

The final stage of an amplifier system is often a 'large-signal' or power amplifier which may be required to supply significant amounts of power to devices such as loudspeakers, motors, relays, etc. These amplifiers generally operate over the whole of the amplifier characteristics and, as a result, tend to introduce some distortion because of the non-linearity of the characteristics at the extremes of operation. Analysis is usually done graphically.

If only a small range of frequencies is required to be amplified, an amplifier can be 'tuned' to these frequencies and will, ideally, not amplify any others. The major use for this type of amplifier is in the radio-frequency amplifier section of communication receivers. Audio frequencies (practically about 40 Hz to 20 kHz), because they are easily absorbed by any object in their path – particularly the earth – are not suitable for direct transmission by radiation from an aerial; instead, they are combined with 'carriers' of much higher frequency. This process of adding or superimposing a low-frequency signal on to a much higher radio-frequency (r.f.) carrier is called 'modulation'. Radio frequencies are usually considered to be those above 100 kHz, although this figure is arbitrary. So, if we transmit the audio-frequency range superimposed on a 100 kHz r.f. carrier, a tuned amplifier will be required to 'select' the range of frequencies – the bandwidth – from 90 kHz to 110 kHz. For commercial reasons – more channels can be allocated without any significant loss in quality of the information transmitted – audio frequencies in the range 400 Hz to 10 kHz are transmitted, giving a bandwidth of about 10 kHz.

9.2 Simple power amplifier

Power amplifiers differ from the small-signal amplifiers considered previously in the method of coupling the load to the collector – the output stage is often required to transfer significant amounts of power. A transformer provides a number of advantages for this purpose. It has a low d.c. resistance, so if there is any quiescent d.c. current it does not consume as much power as would a resistive load. In many instances it is not desirable to pass a d.c. current through the output device, e.g. in the case of a loudspeaker, so a transformer provides isolation. The major advantage of using a transformer, however, is its ability to provide 'impedance matching'. It can be shown that for maximum power transfer between a source and a load, the load and source impedances must be equal. For a transistor in common-emitter mode with a typical output impedance of 2 kΩ and a load such as a loudspeaker, typically 4 to 16 Ω, very little power would be transferred. For a transformer, as shown in fig. 9.1, it can be shown that the effective impedance of the primary, R'_P, as 'seen' by a load connected to the secondary is given by $R'_P = n^2R_L$, where $n = N_1/N_2$, the transformer turns ratio.

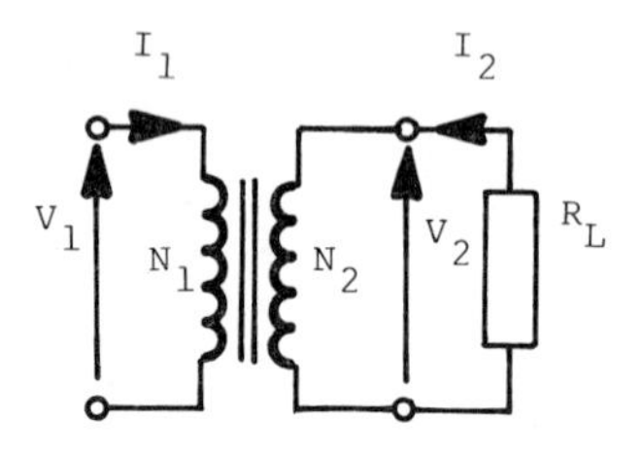

Fig. 9.1 Transformer

Figure 9.2 shows a simple power amplifier operating in class A. Resistors R_1 and R_2 provide d.c. bias; R_E provides thermal stabilisation and is decoupled to a.c. by C_E. The transformer provides the collector load. For d.c. it has negligible resistance; it couples the a.c. signal to the output device while at the same time providing d.c. isolation and, most importantly, providing impedance matching. It is essential that the transistor is adequately rated, i.e. its current and power ratings are adequate.

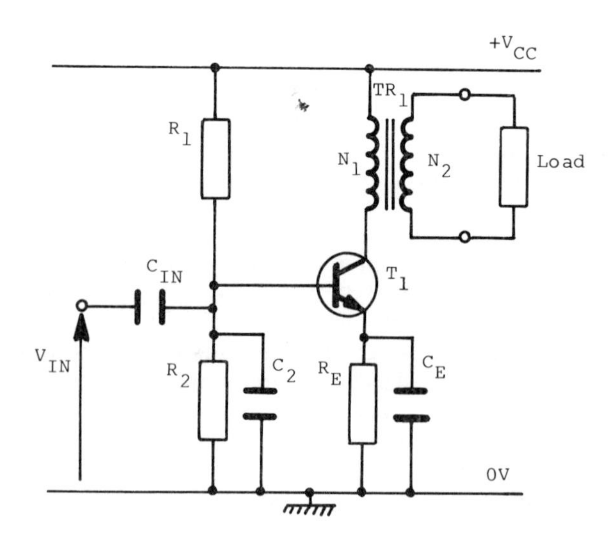

Fig. 9.2 Simple power amplifier

9.3 Push–pull amplifiers

Large-signal amplifiers introduce some distortion due to the non-linearity of the characteristics. This can be reduced to some extent by operating in push–pull. Since a power amplifier operating in class A will dissipate power when no signal is applied, efficiency will be improved if we can operate in class B or class AB.

Figure 9.3 shows a simple push–pull amplifier in class-B bias. Each transistor will be able to handle the same output power as the one in the simple amplifier of fig. 9.2. The transistors should be a matched pair, mounted close together on the same heat sink. Any even harmonics will cancel each other out. (Harmonics are frequencies which are multiples of the wanted frequency. If an amplifier introduces any frequencies which happen to be multiples of any of the signal frequencies, then harmonic distortion will result.) Similarly, any d.c. components in the collector circuit will cancel each other magnetically in the output transformer, preventing saturation.

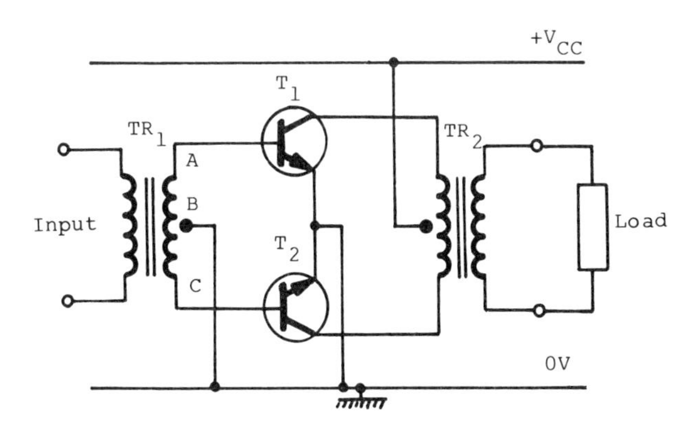

Fig. 9.3 Class-B push–pull stage

Transformer TR_1 is essential to the operation of all push–pull circuits; it is required to 'split' the input signal into two equal but anti-phase signals, so that on each half cycle of the input signal one transistor is driven on and the other off. Transformer TR_2 provides coupling and impedance matching to the load.

If there is no signal, both transistors are off and no current flows in TR_2. If a signal is now applied and on the first half cycle point A of the input transformer is positive with respect to B, B will be positive with respect to C. This means that T_2's base–emitter junction is reversed-biased and T_2 will remain off, but T_1's base–emitter junction is forward-biased and T_1 will start switching on; current will flow in one half of TR_2's primary and will be transformed into the load. On the next half cycle T_1 will be reverse-biased and T_2 driven on; current will flow in the other half of TR_2's primary and the load.

Efficiencies of 70 to 75% are possible with this arrangement. Its major disadvantage is that it introduces serious distortion, known as 'cross-over distortion', due to the non-linearity of the transfer characteristic around the origin. This is very marked with small input signals. The effects of cross-over distortion are shown in fig. 9.4.

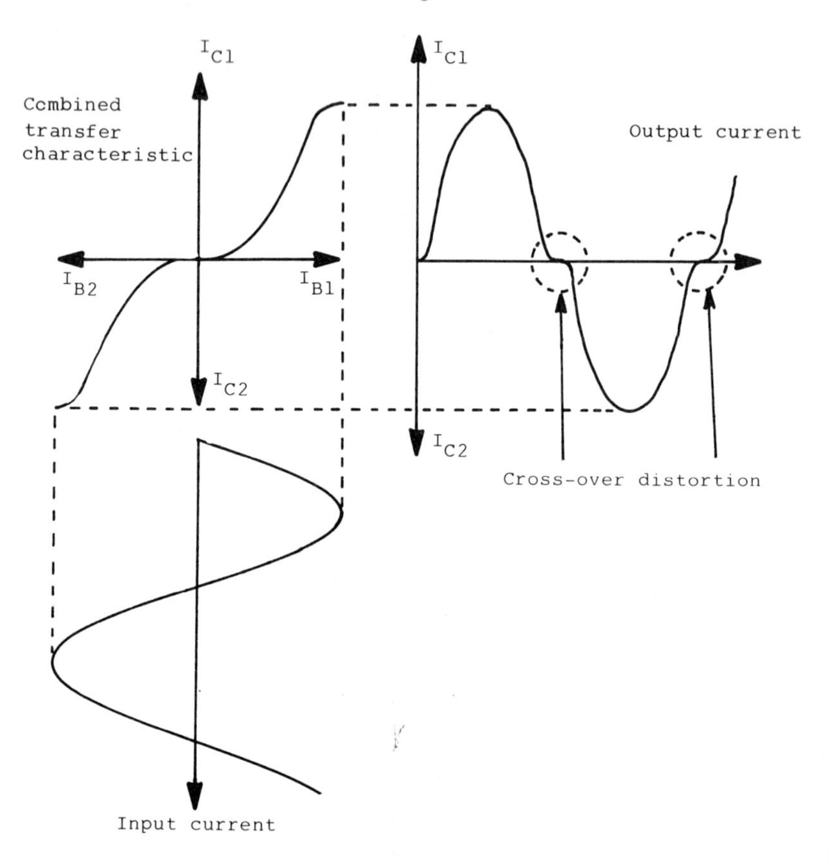

Fig. 9.4 Cross-over distortion

Cross-over distortion can be reduced by biasing the amplifiers into class AB, moving the bias point to produce a transfer characteristic similar to that in fig. 9.5. A small collector current now flows with no signal. A typical circuit for class-AB operation is shown in fig. 9.6. Resistors R_1 and R_2 provide the bias and R_E provides thermal stabilisation; otherwise operation is the same as for the circuit of fig. 9.3.

It is possible to operate push–pull amplifiers in class A, but this offers no advantages over class B or class AB and is less efficient – no more than 50%.

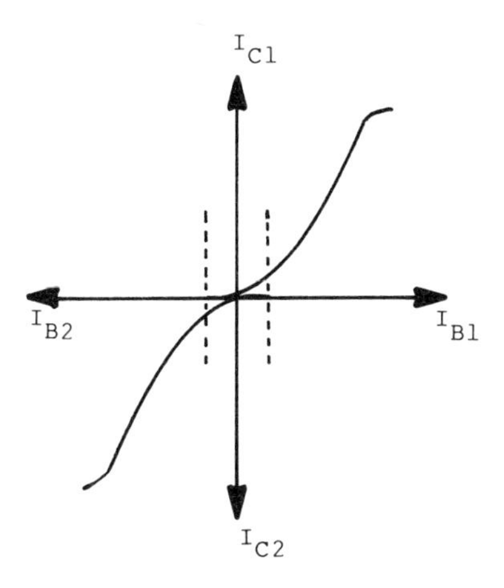

Fig. 9.5 Effect of class-AB bias on combined characteristics

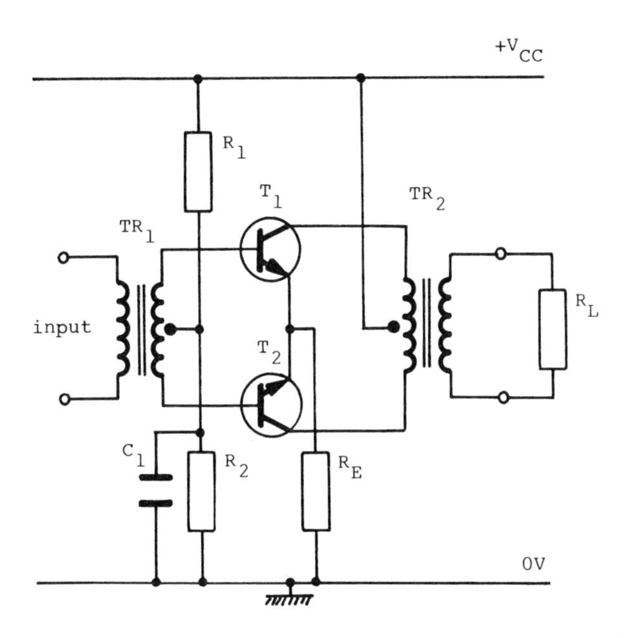

Fig. 9.6 Class-AB push–pull circuit

9.4 Complementary push–pull amplifier

The transistor offers a unique circuit arrangement by combining the properties of an npn transistor and a pnp transistor, i.e. using 'complementary' types. This permits push–pull operation without input or output transformers. The basic circuit arrangement is shown in fig. 9.7. The transistors require opposite polarities for operation.

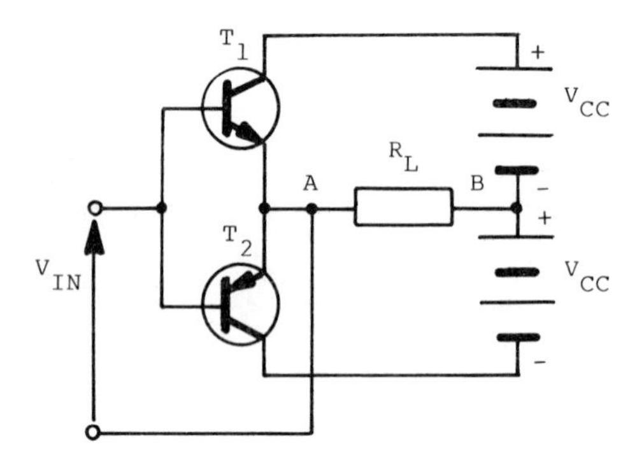

Fig. 9.7 Complementary push–pull circuit

The input signal is fed to the two transistor inputs in parallel. A positive signal results in a collector current flowing in T_1 and current flowing from A to B in R_L. If the input signal is negative, T_2 switches on and its collector current results in current flowing from B to A in R_L. With no signal there is no current in R_L, so the amplifier is operating in class B. To reduce cross-over distortion, class AB would be used. The major advantage of this amplifier is that no isolating capacitor is necessary if the load is a loud-speaker, although in this circuit arrangement it does require positive and negative power supplies.

Matched transistors are essential, though this does increase the cost, since pnp transistors are more expensive because they are harder to make.

9.5 Amplifier stability

Due to the large currents and voltages being used, power stages may go into unwanted or *parasitic* oscillations if there is any unwanted coupling due to poor circuit layout, device capacitances, etc.

These oscillations, which can be any frequency but tend to be above the audio range, are best eliminated in the design stage by ensuring effective isolation between the power and earlier stages – shielding may be used. If oscillations do occur, they can be stopped by changing the active device, by adding 'electronic crutches' such as additional bypass capacitors or r.f. inductors, or possibly by rerouting of the circuit wiring – not easy with printed-circuit boards.

9.6 Industrial power amplifiers

All the previous power amplifiers considered have been 'audio' amplifiers. There are some industrial applications, such as the field control of a Ward–Leonard system (a control system using a constant-speed motor to drive a variable-voltage generator i.e. a motor–generator set), which require signal amplification, but the majority of industrial applications require switching 'on' of power or control of power as discussed in chapter 10 on thyristors.

9.7 Tuned amplifiers

Since the gain of an amplifier depends on the load impedance ($A \approx g_m R_D$ in an FET), if we use a tuned circuit as the collector or drain load we can arrange that the amplifier has a very high gain at some particular frequency and a much smaller gain at all others; thus the amplifier can be used to 'select' a particular frequency, or in practice a small band of frequencies. The 'selectivity characteristic' of a resonant tuned circuit is its ability to discriminate between signals at different frequencies. The selectivity of an amplifier is usually defined in terms of the bandwidth between the 3 dB points. A typical gain/frequency characteristic for a tuned amplifier is shown in fig. 9.8.

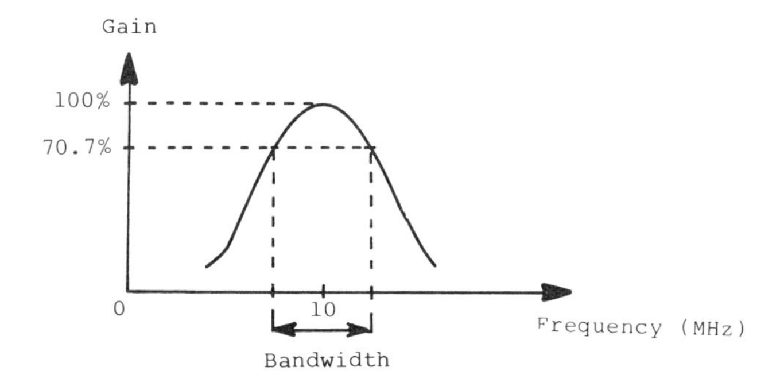

Fig. 9.8 Tuned-circuit characteristic

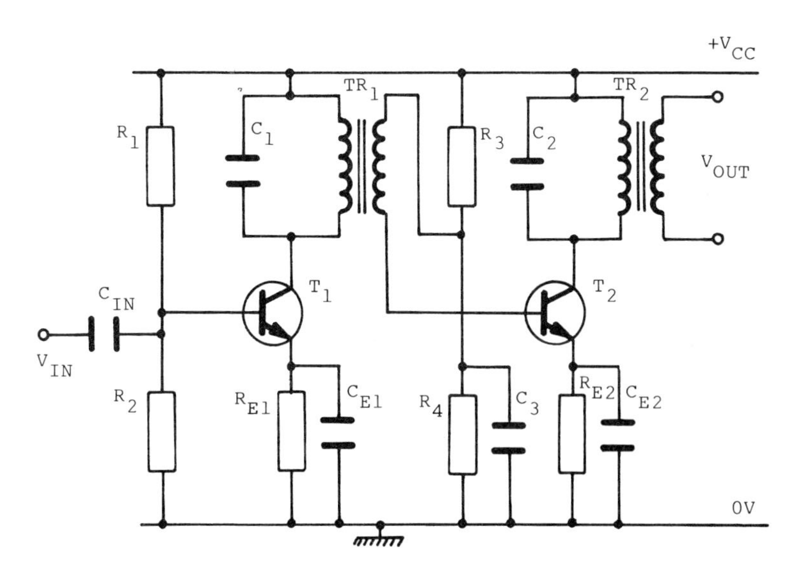

Fig. 9.9 Single-tuned amplifier circuit

Figure 9.9 shows a simple single-tuned amplifier circuit. T_1 and T_2 are connected in common-emitter mode. Resistors R_1, R_2 and R_3, R_4 bias the transistors into class A. Resistors R_{E1} and R_{E2} provide thermal stabilisation, being bypassed for a.c. by C_{E1} and C_{E2}. The collector feeds a single-tuned circuit, with its secondary winding tightly coupled to the tuned-circuit inductance using a step-down ratio to give the correct impedance matching. Because the signal to the second stage is developed across the secondary winding of TR_1, R_4 must be decoupled to a.c. to prevent negative feedback.

For better selectivity a double-tuned circuit would be used, as shown in fig. 9.10. In this circuit, FET's are used as the active devices. Matching is achieved by the relative tapping positions on the primary and secondary of TR_1. The spacing between the coils has to be adjusted to achieve critical coupling to get the required bandwidth.

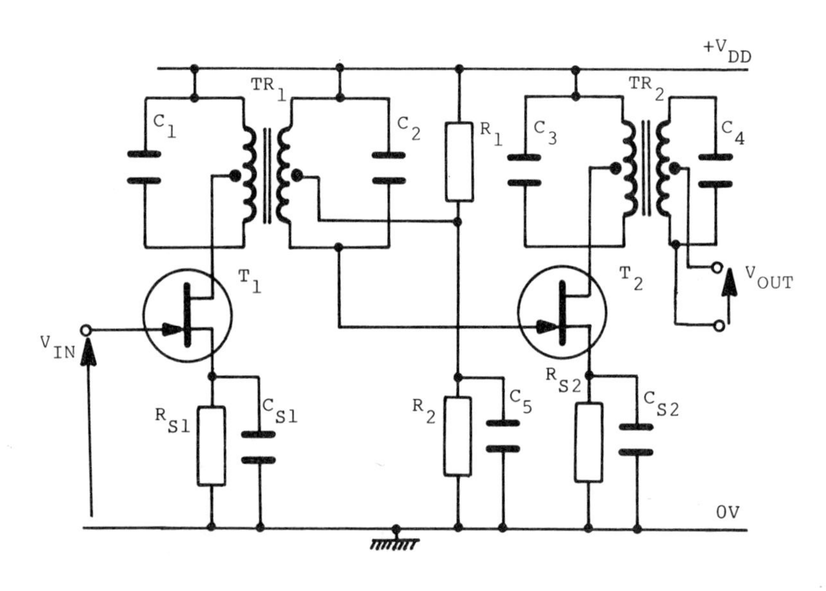

Fig. 9.10 Double-tuned amplifier circuit

Problems

9.1 Why is the analysis of the operation of large-signal amplifiers generally carried out graphically?
9.2 Why is it necessary to use a radio-frequency 'carrier' to transmit audio frequencies?
9.3 How can the radio-frequency carrier be 'selected' by an amplifier?
9.4 What are the advantages of using a transformer to couple the load in the final stage of an amplifier?

9.5 What are the disadvantages of operating a power amplifier in class-A bias?
9.6 Explain the term ‘matched’ when applied to transistors.
9.7 What disadvantages result from operating a push–pull amplifier in class-B bias?
9.8 What is the function of the input transformer in a push–pull amplifier?
9.9 What are the causes of parasitic oscillations in amplifiers? How can they be reduced?
9.10 How is cross-over distortion reduced in push–pull amplifiers?

10 Thyristors

10.1 Introduction

The 'thyristor' family of semiconductor devices includes any four-layer p–n–p–n device, but it is the 'reverse blocking triode thyristor' or *silicon controlled rectifier* (SCR) which most people mean when referring to the 'thyristor'. Other devices in the family are the diac, triac, silicon unilateral switch, silicon bilateral switch, programmable unijunction transistor, and gate turn-off thyristor.

The main thyristor applications involve control or conversion of power. Modern thyristors can be made to handle continuous current in excess of 2.5 kA.

10.2 The silicon controlled rectifier (SCR)

The SCR is basically a power diode whose periods of conduction can be controlled by an external signal. It has three terminals, called anode, cathode, and gate. The symbol and simplified structure are shown in fig. 10.1.

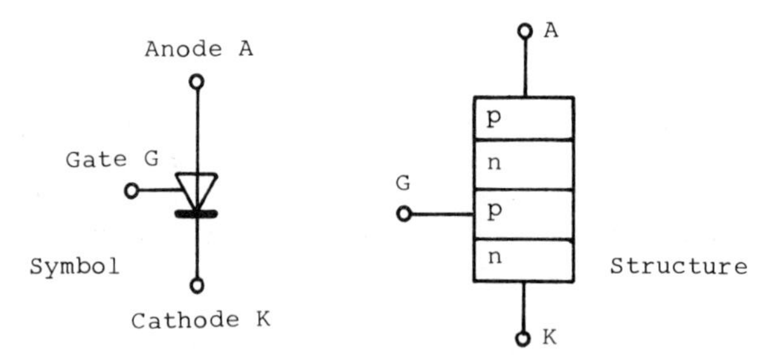

Fig. 10.1 SCR symbol and structure

10.2.1 Theory of operation

With no potentials applied to the SCR there will be three p–n junctions – J_1, J_2, and J_3 – formed as shown in fig. 10.2(a). If the anode is made negative with respect to the cathode, J_1 and J_3 will be reverse-biased and widened but J_2 will be forward-biased, as shown in fig. 10.2(b). Only a small thermally generated leakage current will flow, and this can be ignored for all practical purposes. The SCR is now said to be 'reverse-blocking'. The device will not normally conduct in this direction.

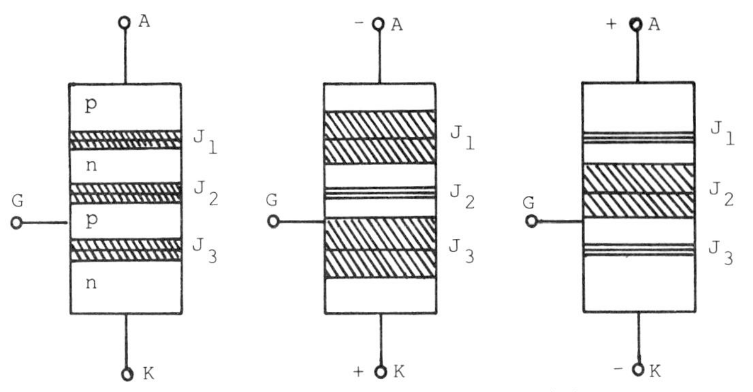

Fig. 10.2 SCR depletion regions

If the anode is now made positive with respect to the cathode, J_1 and J_3 will be forward-biased and become narrower but J_2 will be reverse-biased and wider and again only a small leakage current will flow – this is shown in fig. 10.2(c). The SCR is now said to be 'forward-blocking'. If the gate is now made positive and a current above a minimum specified value flows, the SCR will turn on and current flows from anode to cathode as in an ordinary diode. After a short time, the gate current can be reduced to zero without affecting the anode current – the gate loses all control and cannot be used to switch the device off. The SCR is now said to be 'forward-conducting' – it will conduct as long as the anode current is above its 'holding' value. How a gate current causes the SCR to turn on can best be explained by considering the SCR using a 'two-transistor' model.

Figure 10.3 shows the simple 'two-transistor' model of the SCR. We can consider it to be a pnp transistor T_1 and an npn transistor T_2, each of which is biased in the common-emitter mode. The base, emitter, and collector

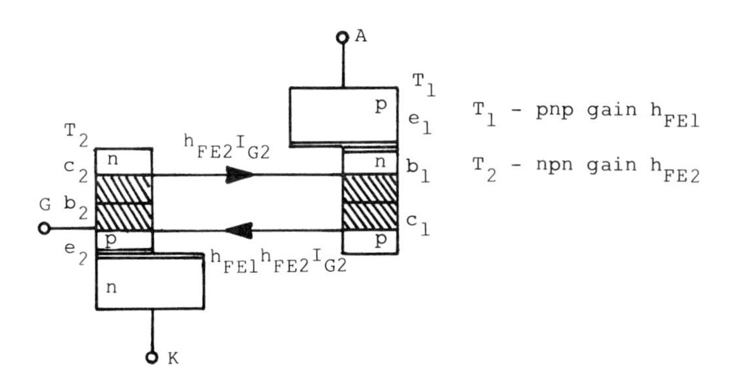

Fig. 10.3 'Two-transistor' model of the SCR

for each are shown on the diagram. The gate current is the base current of T_2, I_{G2}. This will be amplified by normal transistor action to give a collector current of $h_{FE2}I_{G2}$ which forms the base current of T_1; this is amplified by T_1 to give it a collector current of $I_{G2}h_{FE2}h_{FE1}$ which then flows back into the base of T_2, forming the gate current. If $I_{G2}h_{FE2}h_{FE1} > I_{G2}$ then the action will be regenerative (i.e. the current is amplified and fed back and amplified again etc.) and cumulative – once started, it continues rapidly until the depletion region at J_2 is destroyed and the anode–cathode resistance becomes very low and a high current flows. The SCR can only be turned off by reducing the anode current below its holding value – this is usually done by reducing the anode voltage to zero.

Figure 10.4 shows the anode characteristics. If the forward voltage is increased above the forward breakover voltage V_{BO}, the SCR will avalanche into conduction. Normal operation is with the anode voltage held below V_{BO} and a trigger pulse, I_{GT}, applied to the gate as shown.*

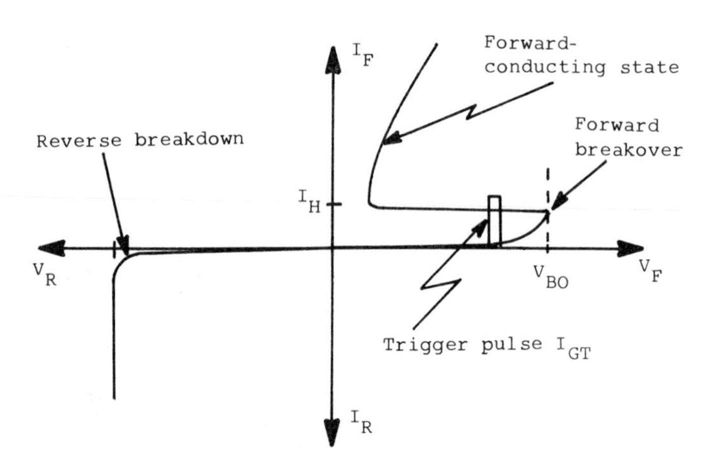

Fig. 10.4 SCR characteristics

The gate current required to trigger the device is very small compared to the anode current the devices are capable of handling; e.g. for a BTW23 series device the average r.m.s. on-state current is 140 A maximum and the device will trigger if $I_{GT} > 150$ mA. The forward volt drop across a conducting SCR is about 1 V; hence, if the current is 100 A, the SCR will have to dissipate about 100 W – this may require the use of heat sinks.

* Some textbooks show characteristics which imply that lower gate currents are required to trigger the SCR for higher values of anode voltage. If there is *any* difference, it is undetectable under normal circumstance with commercially available devices.

10.3 Thyristor firing techniques

There are three basic thyristor firing techniques. Each has its own merits, and they are discussed below.

10.3.1 Phase control

This is the simplest and cheapest method of thyristor firing. The trigger current is delayed for part of the supply-voltage period – this is the trigger angle α. The time the current flows, and hence the power consumed by the load, is controlled and is proportional to the conduction angle θ. If the device is operating on a.c., it will automatically switch off at the end of each half cycle as the supply reverses – current can flow only during the positive half cycle. Figure 10.5 shows typical waveforms.

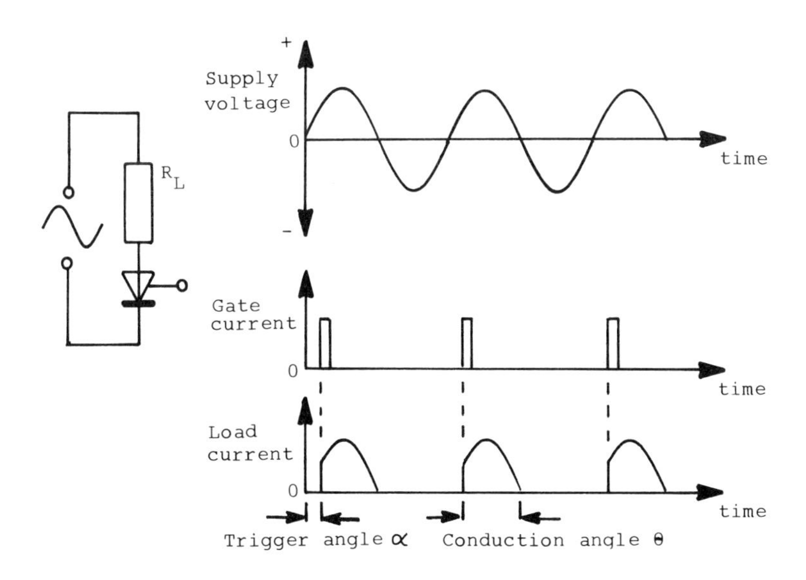

Fig. 10.5 SCR waveforms

Two methods of achieving phase control are shown in fig. 10.6. The purely resistive method in fig. 10.6(a) only permits control up to 90°. The capacitor in fig. 10.6(b) provides an extra 90° phase shift, permitting control over the full 180°. Phase control is used where simple inexpensive but limited control is required.

10.3.2 Pulse firing

Applying a pulse to the gate of the SCR will cause it to trigger. This offers a number of advantages over the method in section 10.3.1:

a) the SCR can be fired anywhere in the positive half cycle;
b) the gate circuit can be coupled to the SCR by transformer, so that:

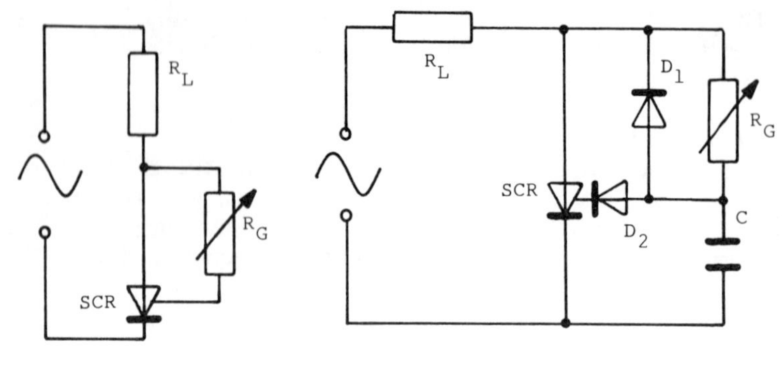

Fig. 10.6 Phase-control methods

i) electrical isolation is provided between power and control circuits,
ii) more than one gate can be fed simultaneously,
iii) pulses of one polarity can be used to fire one SCR while pulses of the other polarity can fire another;

c) high pulse powers can be used to ensure that the SCR fires under all conditions;
d) a chain of pulses may be generated and synchronised to the anode supply when a continous firing signal is required;
e) pulses have fast rise times, making them ideal waveforms for triggering SCR's.

Often the pulses are fired in 'bursts', a technique known as 'burst firing'.

10.3.3 Zero-voltage switching

Both the above methods introduce mains-borne and radiated r.f. interference due to switching-on part way through a cycle, particularly when controlling large powers, since large transients (voltage spikes) are generated on the power-supply lines.

To overcome these problems, in 'zero-voltage switching' the SCR is switched only as the supply changes from negative to positive at the beginning of the cycle. To achieve power control, the device will be off for complete cycles; i.e. control is effected by 'skipping' whole cycles. Commercially available 'zero-voltage switches' are used to achieve the control. This technique significantly reduces radio-frequency interference.

10.4 Thyristor protection

Thyristors are particularly prone to false triggering and/or catastrophic failure when subjected to high rates of change of voltage or current with respect to time (high dv/dt and di/dt) and transient voltage spikes.

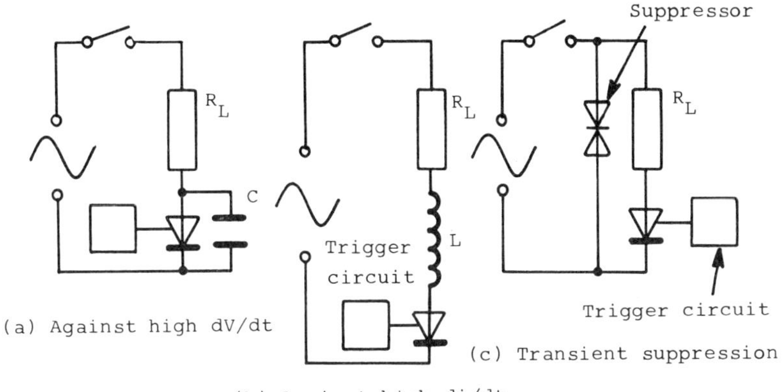

Fig. 10.7 Thyristor protection

Protection against high dv/dt is achieved using a resistor–capacitor network connected across the SCR to limit the rate of rise of voltage – this is shown in fig. 10.7(a). A major application which can cause a high dv/dt to be generated is when controlling inductive loads, e.g. motors – when the supply reverses, the current can be lagging the supply by a considerable angle and attempting to force the current to zero quickly results in very high rates of dv/dt. Networks to prevent this are usually referred to as 'snubbers'.

Reduction of high di/dt can be achieved using an inductor as shown in fig. 10.7(b).

Transient suppression can be achieved using voltage-dependent resistors, neon lamps, or inverse-breakdown diodes (diodes used in reverse bias across a d.c. or a.c. supply which only breakdown at voltages higher than normal, 'clamping' the voltage at the breakdown value), shown in fig. 10.7(c).

10.5 The triac (bi-directional triode thyristor)

For general use on a.c., two SCR's in 'inverse parallel' are required, as shown in fig. 10.8.

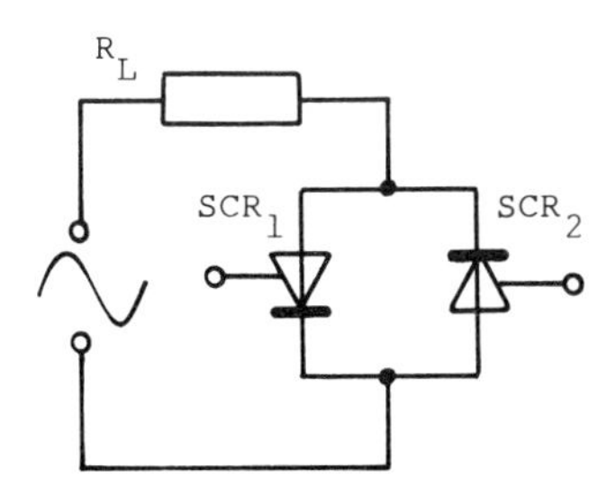

Fig. 10.8 SCR's in inverse parallel

The triac is a single three-terminal device which is equivalent to the circuit of fig. 10.8. It can conduct or block in either direction and can be triggered on with either polarity of gate signal.

The internal construction is more complicated than for the SCR, but its triggering methods, applications, characteristics, ratings, and protection methods are similar. Figure 10.9 shows the triac symbol and characteristics. A simple triac circuit and associated waveforms are shown in fig. 10.10.

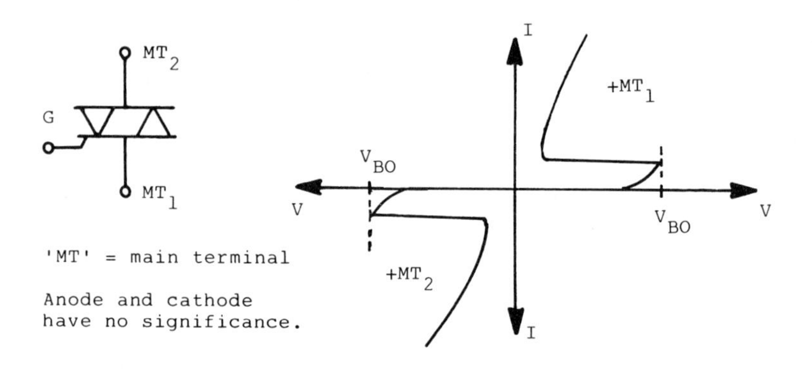

Fig. 10.9 Triac symbol and characteristic

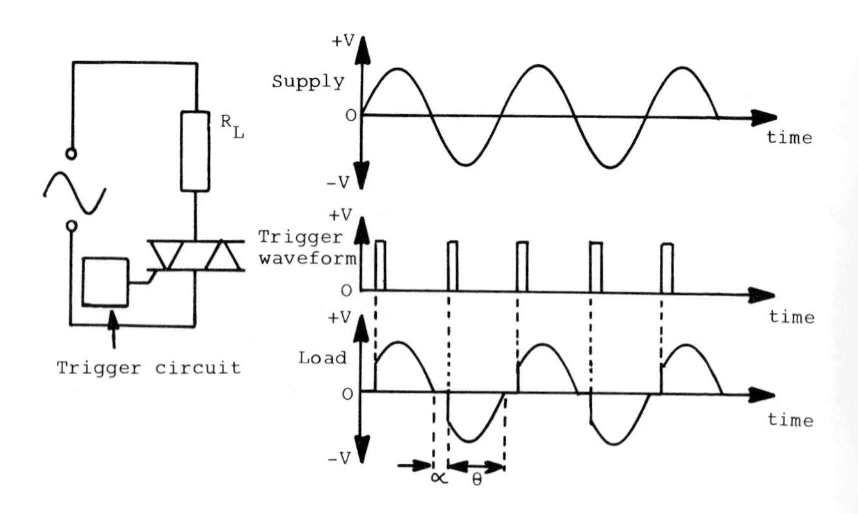

Fig. 10.10 Simple triac circuit and waveforms

10.6 The diac (bi-directional diode thyristor)

The diac is used mainly as a trigger device for SCR's and triacs. It has essentially a transistor structure with a negative resistance characteristic once its breakover voltage, V_{BO}, is exceeded. Breakdown occurs with either polarity of signal at about 32 V. Provided the current is above the minimum switching value, I_{BO}, the device will remain on. Figure 10.11 shows the symbol and characteristics.

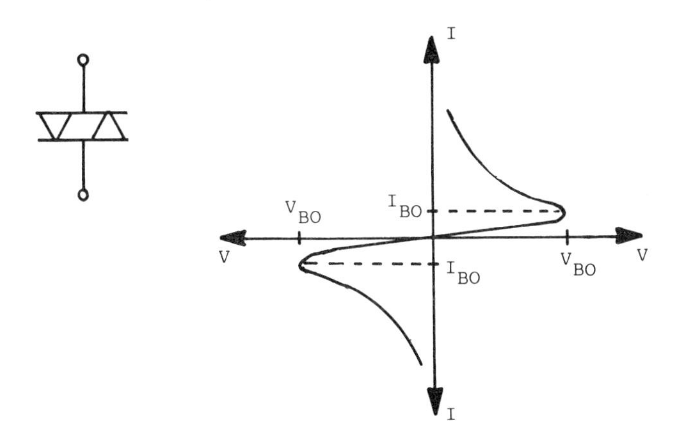

Fig. 10.11 Diac symbol and characteristics

Figure 10.12 shows a simple triac trigger circuit using a diac. When capacitor C_1 has charged to about ± 32 V, the diac will breakover and the capacitor will discharge through the diac, providing a trigger pulse to the triac allowing current to flow in the load. Increasing the resistance R_1 will delay the time at which V_{BO} is reached and thus when the load is switched on.

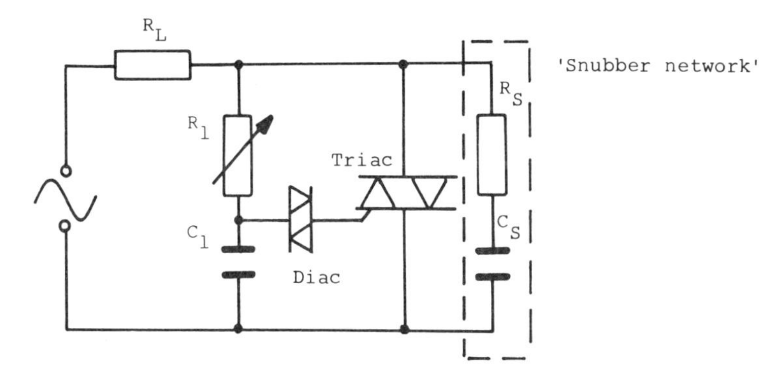

Fig. 10.12 Simple diac trigger circuit

As with SCR's, high dv/dt must be avoided but, since the triac can conduct in either direction, it is essential to ensure that current has fallen below its holding value *before* voltage reversal takes place. When used with inductive loads, 'snubber' networks are essential.

10.7 Silicon unilateral switch (SUS)

This is basically an SCR with the gate connected to the anode and an internal low-voltage avalanche diode between the gate and the cathode. Figure 10.13 shows the symbol and characteristics.

The SUS is usually used as a relaxation oscillator (explained in chapter 12) in trigger circuits for SCR's and triacs.

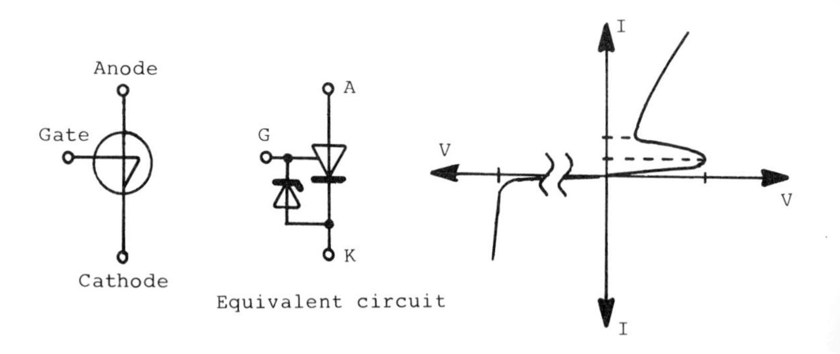

Fig. 10.13 SUS symbol and characteristic

10.8 Silicon bilateral switch (SBS)

This is essentially two SUS's in inverse parallel, as shown in the equivalent circuit of fig. 10.14. The symbol and characteristics are also shown. As it can be triggered with either polarity of signal, the SBS is particularly useful for triggering triacs.

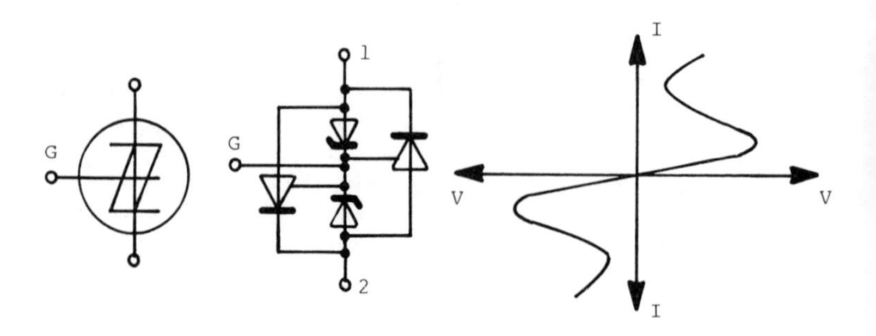

Fig. 10.14 SBS symbol and characteristic

10.9 Unijunction transistor (UJT)

This is not a thyristor device but is commonly used in thyristor trigger circuits. It is a three-terminal device. Figure 10.15 shows the structure, symbol, and characteristics.

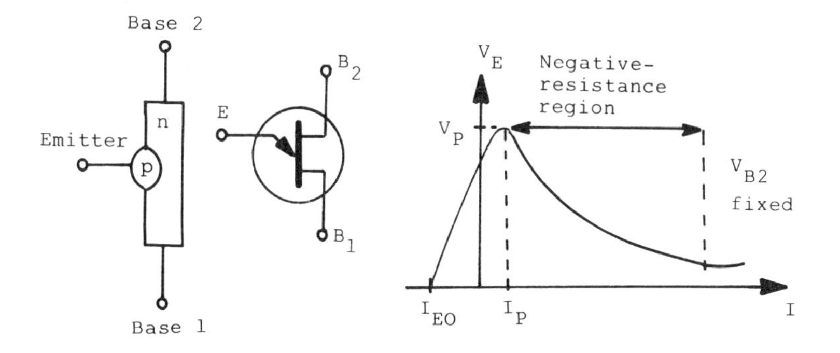

Fig. 10.15 UJT structure, symbol, and characteristic

The device consists of a lightly doped ‘n’ bar and a heavily doped p-type emitter region. Between the terminals B_1 and B_2 the device is a simple resistor whose value is determined by the dimensions and doping level.

If the emitter voltage V_E is less than the trigger voltage V_P, the emitter junction is reverse-biased and only the leakage current I_{EO} flows. As V_E rises above V_P, the emitter current rises above the peak current I_P and the UJT turns on. The voltage V_P is determined during manufacture by the relative position of the emitter on the bar. In the on condition, the resistance between E and B_1 is very low, and any current flow will be limited only by any external resistance.

Figure 10.16 shows the basic relaxation oscillator, and waveforms at E and B_1, using a UJT.

The capacitor is charged through the resistor R until the voltage at the emitter reaches V_P, when the UJT will turn on, discharging the capacitor rapidly through the low-value resistor R_{B1}. As the capacitor discharges, the emitter voltage falls; when it reaches about 2 V the emitter ceases to conduct and UJT turns off and the cycle is repeated. The frequency of oscillation is determined by the R–C time constant and is approximately $1/RC$; it is virtually independent of supply voltage and temperature.

10.10 Programmable unijunction transistor (PUJT)

This is a thyristor device with characteristics similar to those of the UJT. Instead of using the internal device resistances to determine the trigger voltage V_P (this is the voltage at which the device resistance falls to a low value), external resistors (R_1 and R_2 in fig. 10.17) are used. The trigger voltage V_P is determined by the choice of R_1 and R_2, providing the device's programmability.

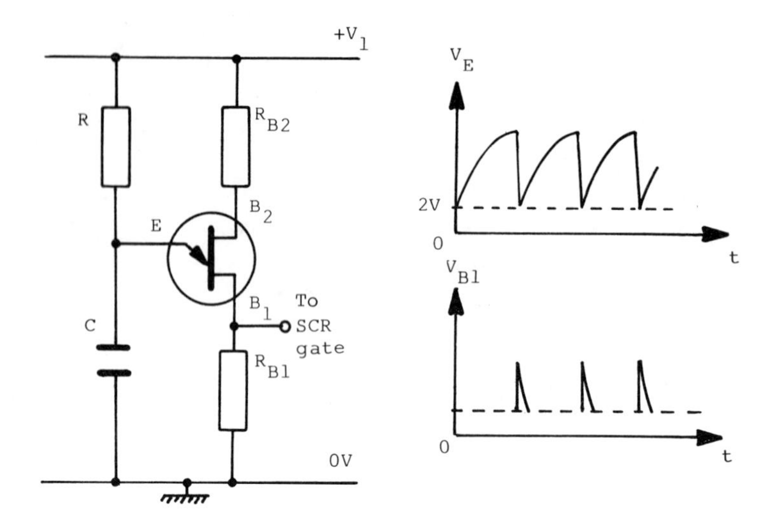

Fig. 10.16 Basic UJT relaxation oscillator

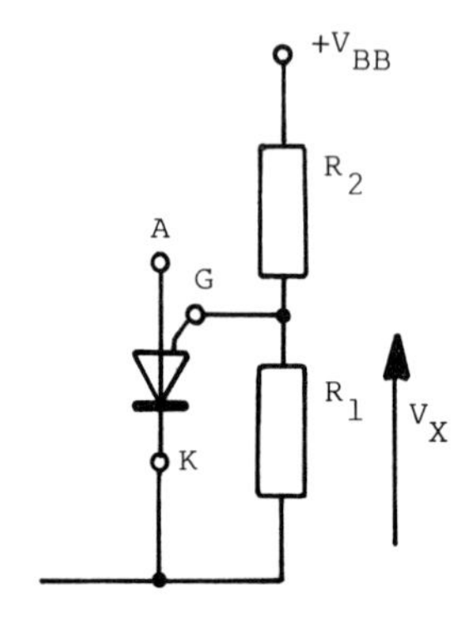

Fig. 10.17 Basic PUJT circuit

PUJT's would be used to provide trigger pulses where long time intervals are required between pulses.

10.11 Gate turn-off thyristor

This device is similar to the SCR but can be turned off by applying a negative gate signal. It is not particularly efficient, because it requires up to 20 times the turn-on voltage to turn it off. Figure 10.18 shows the symbol.

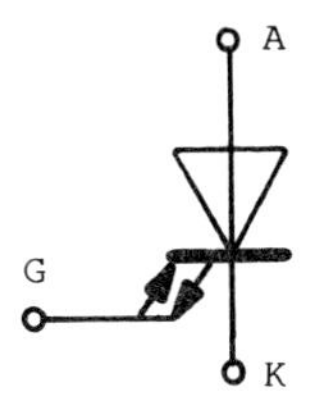

Fig. 10.18 Gate turn-off SCR symbol

10.12 Thyristor applications

There are numerous applications for thyristors. Three simple examples are given below.

10.12.1 Simple battery charger

Figure 10.19 shows a practical battery-charger circuit. The zener diode holds the voltage at point 'X' at 15 V. If the battery is discharged, the voltage between gate and cathode, V_{GK}, is sufficiently positive to switch the SCR on. As the battery charges, its terminal p.d. rises, reducing V_{GK} ($= V_D - 0.6\text{ V} - V_{BAT}$), raising the cathode voltage above the anode, and turning the SCR off. D_1 blocks the SCR discharge path.

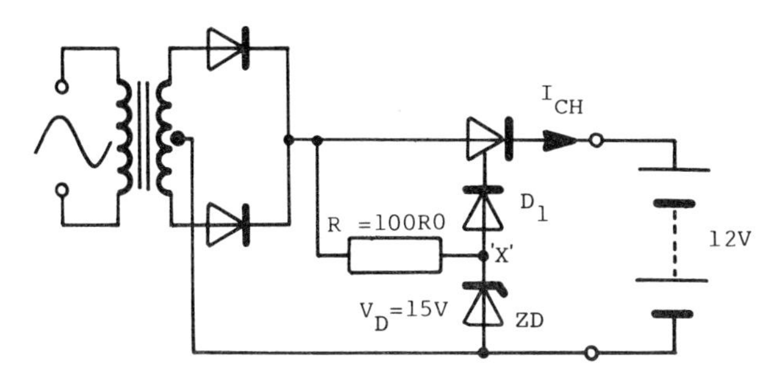

Fig. 10.19 Simple battery charger

10.12.2 Overvoltage protection

Considerable damage can be done to many modern circuits if excessive voltage is applied; for example, TTL devices must not have more than 7 V applied to them. The simple 'electronic crowbar' of fig. 10.20 is a typical overvoltage-protection circuit. The zener breakdown voltage is chosen to be slightly higher than the maximum acceptable value of V_{IN}. If V_{IN} rises above an acceptable level, the zener starts to conduct and the volt drop across resistor R – the SCR gate voltage – rises and will switch the SCR on, putting a short across the terminals and blowing the fuse. The capacitor C prevents transients causing spurious operation, particularly on switch-on.

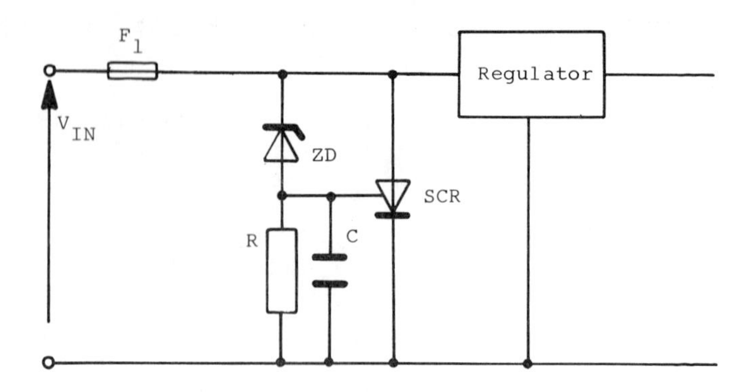

Fig. 10.20 'Electronic crowbar'

The SCR must have a rating considerably higher than the fuse, so that the SCR does not 'protect' the fuse.

10.12.3 Motor speed control

Figure 10.21 shows a typical induction-motor speed control which provides continuous control for washing machines etc. The triac controls power to the 'run' winding and hence speed. Control could be provided by a UJT oscillator whose supply could be obtained via a tachometer coil (a coil usually with a permanent magnet mounted inside – the magnet is rotated on a drive shaft) on the motor shaft. Since the output from the tachometer is proportional to speed, it would then be used to maintain the desired speed setting.

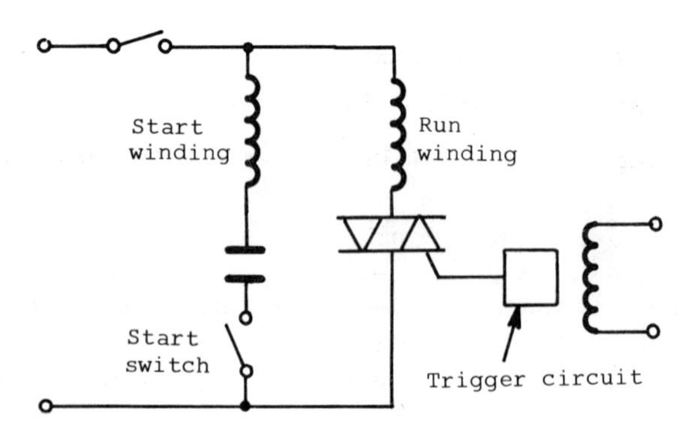

Fig. 10.21 Induction-motor speed control

Example 10.1 An SCR is connected in series with a 240 V 60 W lamp to the mains supply. The supply is also connected via a 17 kΩ resistor to the gate. Conduction will start in the device when the gate current reaches 10 mA. Assuming that the internal resistance of the SCR can be neglected when in the 'on' state, determine the average current in the lamp.

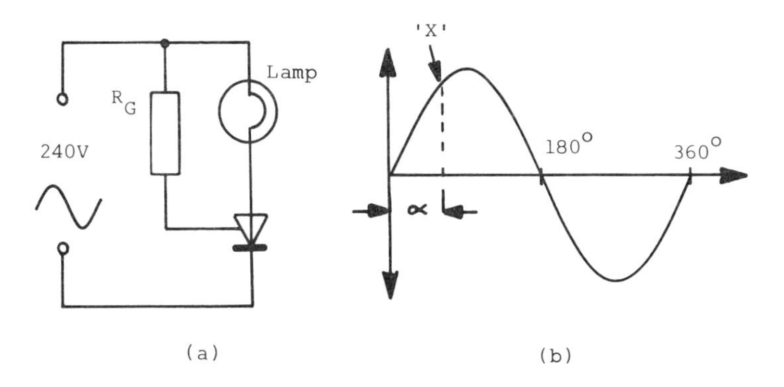

Fig. 10.22 Circuit and waveform for example 10.1

Figure 10.22(a) shows the circuit arrangement.

Since the SCR requires 10 mA to trigger it,

$$\text{gate trigger voltage } V_{GT} = I_{GT} \times R_G = 10 \text{ mA} \times 17 \text{ k}\Omega = 170 \text{ V}$$

The voltage will reach 170 V – point 'X' in fig. 10.22(b) – after angle α. Angle α can be found from

$$V_{MAX} \sin \alpha = 170 \text{ V}$$

where $V_{MAX} = \dfrac{V_{RMS}}{0.707} = \dfrac{240}{0.707} = 339.5 \text{ V}$

hence the trigger angle α is given by

$$\alpha = \arcsin(170/339.5) = 30° = \pi/6 \text{ rad}$$

The average voltage across the lamp can then be found from

$$V_{AV} = \frac{\int_{\pi/6}^{\pi} V_{MAX} \sin \theta \, d\theta}{2\pi}$$

$$= \frac{V_{MAX} \Big[-\cos\theta \Big]_{\pi/6}^{\pi}}{2\pi} = \frac{V_{MAX}[-\cos\pi + \cos(\pi/6)]}{2\pi}$$

$$= \frac{339.5 \text{ V} \times [-(-1) + 0.866]}{2\pi}$$

$$= 339.5 \text{ V} \times 1.866/2\pi$$

$$= 100.8 \text{ V}$$

$$\text{Average current } I_{AV} = \frac{V_{AV}}{R_L}$$

where R_L is the resistance of the lamp, given by

$$R_L = \frac{V^2}{P} = \frac{(240 \text{ V})^2}{60 \text{ W}} = 960 \ \Omega$$

$$\therefore \ I_{AV} = \frac{100.8 \text{ V}}{960 \ \Omega} = 105 \text{ mA}$$

Problems

10.1 An SCR is connected in series with a 560 Ω load across a 240 V mains supply. Calculate the average value of the load current if the SCR is triggered at 30° on each positive half cycle of the supply. (Assume that the SCR is ideal.)

10.2 The SCR in fig. 10.23 starts conducting at a gate current of 10 mA. Assuming ideal SCR characteristics, calculate

a) the angle at which conduction begins,

b) the average current in the load.

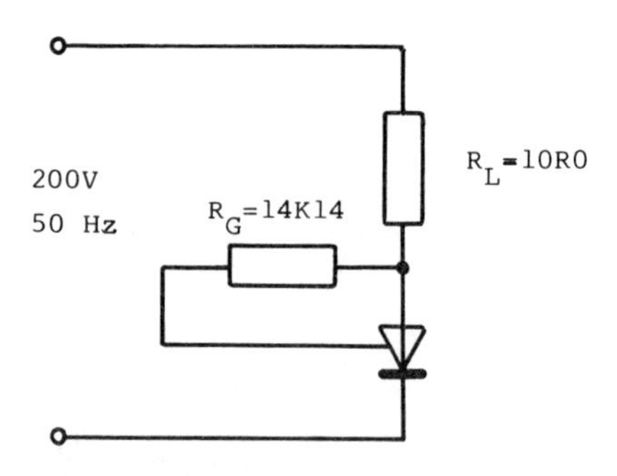

Fig. 10.23 Circuit for problem 10.2

10.3 Assuming that the supply voltage V to the SCR trigger circuit in fig. 10.24 is present only from the beginning of each cycle until the SCR triggers, if the SCR triggers when the UJT emitter voltage V_E ($= V_C$) reaches 2 V, calculate

a) the trigger angle α (the angle at which the SCR switches on),

b) the average power developed in the lamp,

c) a suitable power rating for the SCR.

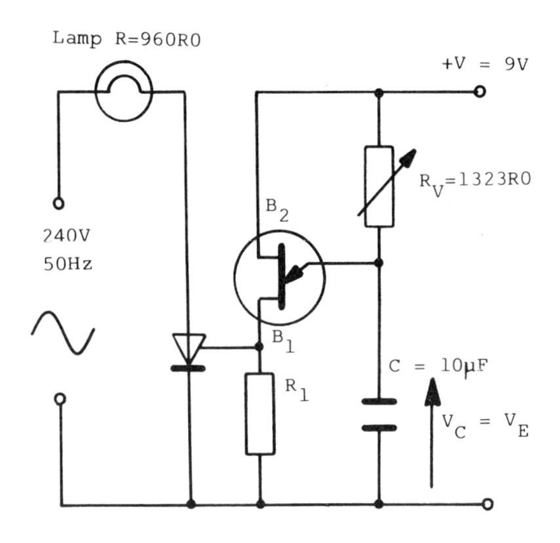

Fig. 10.24 Circuit for problem 10.3

10.4 The average forward volt drop across a conducting SCR will be approximately
a) 600 V.
b) 20 V.
c) 1 V.
d) 0.2 V.
e) 250 V.
10.5 What type of device is each of the following: (a) BTW34, (b) BT138, (c) BTX94?
10.6 For the BTX94, determine the r.m.s. current which the device can handle: $I_{T(RMS)}$ = ... Hence calculate the approximate power that the device will have to dissipate if it is being run continuously.
10.7 Explain why 'zero-voltage switching' is preferred for SCR's and triacs.
10.8 Use the 'two-transistor' model to explain the operation of the SCR.
10.9 Explain what protection must be provided, and why, to prevent SCR's and triacs triggering when no gate signal has been applied.
10.10 What similarities exist between all thyristor devices?

11 Integrated circuits

11.1 Introduction

Integrated circuits (IC's) are complete electronic circuits containing resistors, capacitors, diodes, transistors, and their associated electrical interconnections manufactured on a single chip of silicon. Figure 11.1 shows typical packages used for IC's.

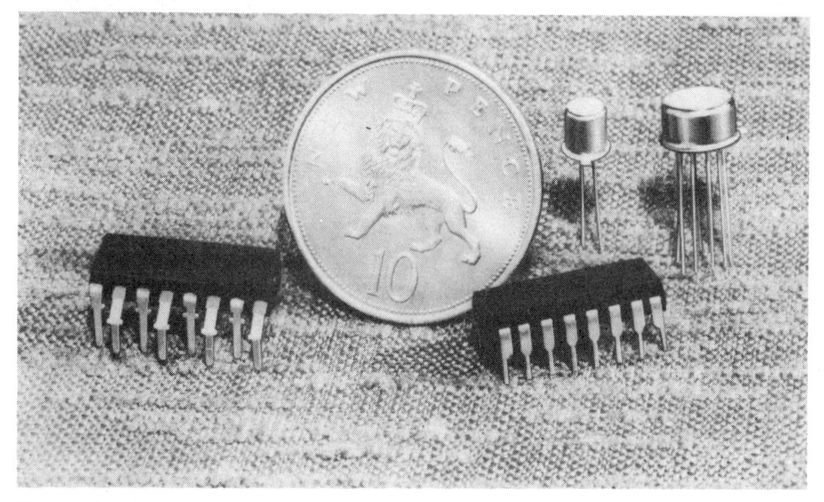

Fig. 11.1 IC packages (coin diameter 28 mm)

The active part of the device is a tiny chip of silicon, which may be no bigger than that for a single discrete transistor such as a BC109. Devices such as microprocessors do come in larger sizes, but most of the increased size is necessary to enable connections to be made to the outside world. The connecting leads must be rigid enough to permit them to be inserted into IC holders. The complete package is extremely rugged.

A typical chip is shown greatly magnified in fig. 11.2. This is a very simple chip – most IC chips are bigger. Some of the components are identified in fig. 11.2: T = transistor, R = resistor, D = diode.

11.2 Advantages of IC's

The electrical circuitry within the IC could be constructed with discrete components, but IC's have many advantages over conventional discrete circuits – e.g. small size (miniaturisation), cheapness, reliability, low

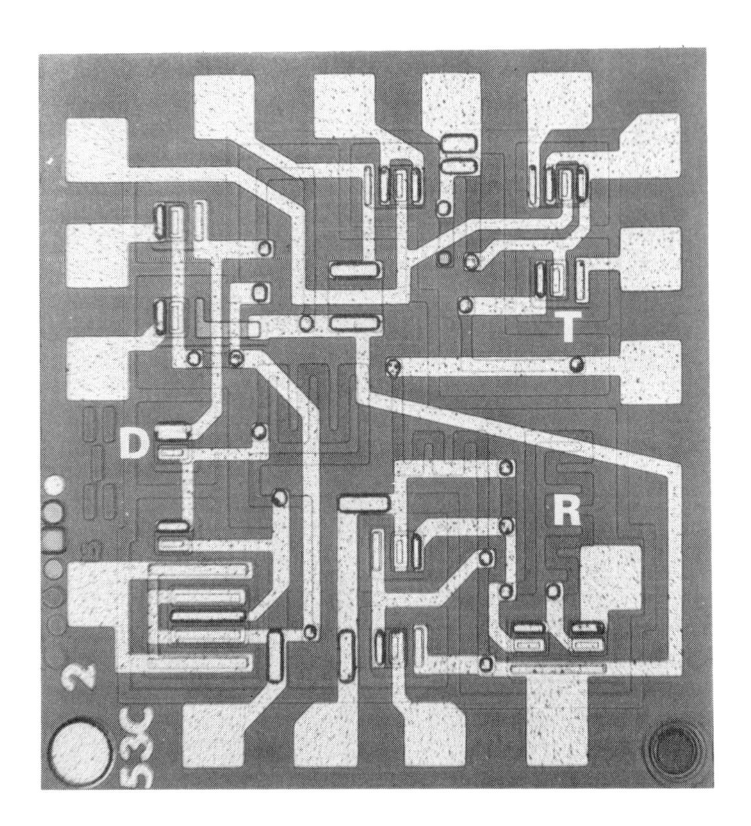

Fig. 11.2 IC chip

power consumption, standard packaging, easy replacement with devices with similar specifications, standard power-supply requirements, reduced interference between components on the chip (due to their close proximity and the reduced number of leads), easy extension of internal circuitry (by the manufacturer), reduced effects of temperature on drift (due to the close proximity of components), and better matching of parameters.

11.3 Limitations of IC's

The major limitation of the IC is its inability to dissipate large amounts of power, due to its small size. Any increase in current produces heat which may destroy the device by melting the active regions. Thus IC's are usually limited to information processing, and the 'working' or output stages are usually discrete components.

Due to the small physical size of IC devices, the separation between conductors is very small, restricting the voltages which may be applied.

Manufacturing techniques do not permit accurate control over component values, and at present inductors and transformers cannot be produced on IC's.

11.4 Manufacture of integrated circuits

IC's are produced hundreds at a time on a slice of pure silicon about 5 cm in diameter. Once the process is complete, the wafer is cut up into individual circuits which are then mounted and encapsulated.

The individual components and circuits are built up in layers, producing a three-dimensional circuit.

The 'artwork' for each layer is produced using an accurately made 'photo mask' and is then reduced by photographic means for the entire slice with hundreds of identical circuits on it.

After each process step, the slice is left with a layer of oxide over its surface, and 'windows' are etched into this layer to permit the subsequent introduction of impurities into the slice. The sequence of operations is as follows. A compound called 'photoresist' is applied to the slice to form a thin even coating. The mask, consisting of a pattern of opaque areas on a glass sheet, is placed on the slice and ultra-violet light is shone through it. This has the effect of hardening the exposed areas of photoresist. The soft unexposed areas can then be removed by a solvent. Then, by etching with an acid, the exposed oxide areas can be selectively dissolved. Following this step, the remaining photoresist is removed. The pattern of the mask has now been transferred to the slice.

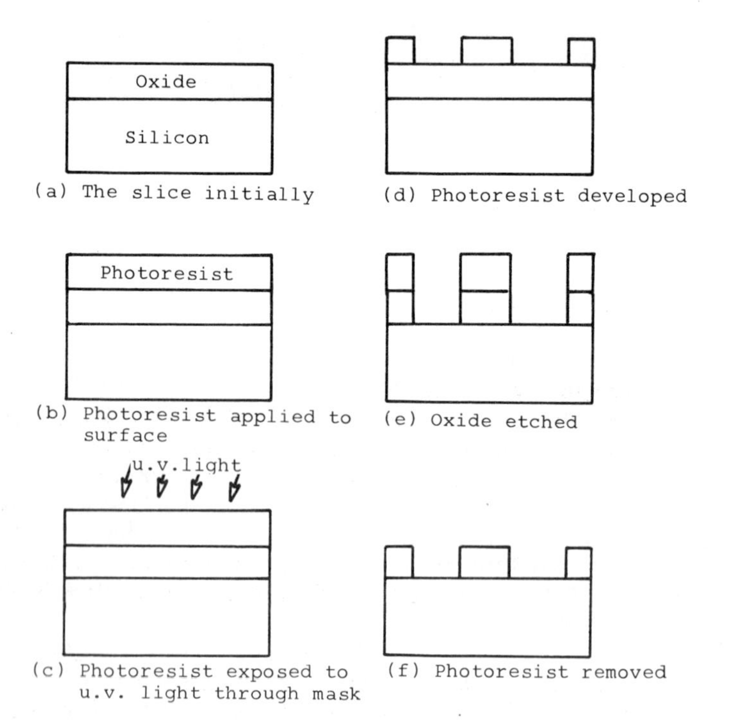

Fig. 11.3 IC etching sequence

Once the windows have been formed in the oxide, the slice is placed in a diffusion furnace and heated to about 1200°C in an atmosphere containing a suitable donor impurity. The donor penetration depth is determined by the length of time the slice is exposed in the furnace.

These intricate steps are repeated for each successive layer required until the whole circuit has been fabricated.

The window-etching sequence is shown diagramatically in fig. 11.3, and fig. 11.4 shows a typical slice after processing.

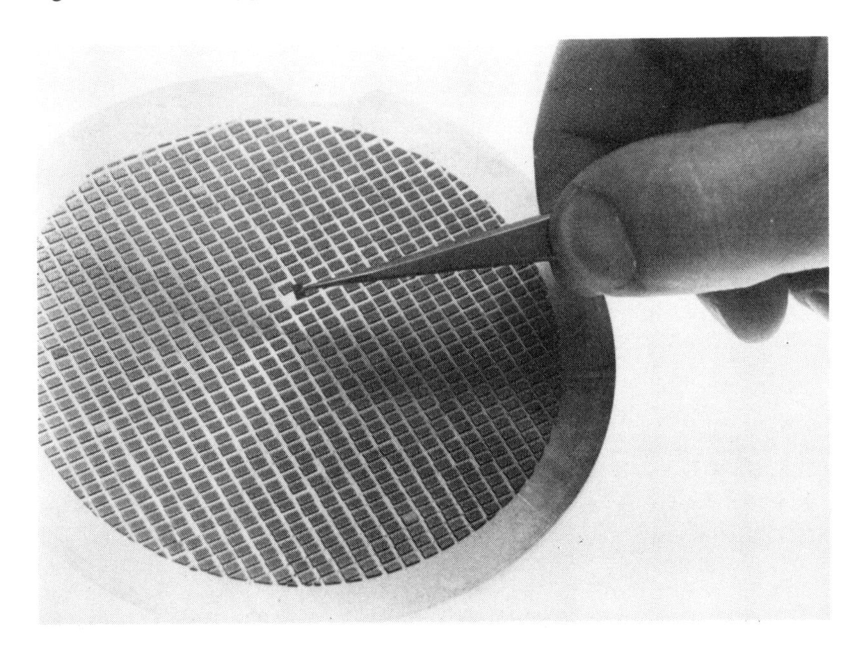

Fig. 11.4 Processed IC slice

11.4.1 Mounting and bonding

Once its processing is complete, the slice is cut up into individual circuits and the electrical connections are made by vibration welding. The IC is then encapsulated in a suitable package to provide protection.

11.4.2 Testing integrated circuits

After assembly, each IC is put through a series of tests of its electrical performance to make sure it meets the data-sheet specifications over its working temperature range.

IC's are usually tested first in the range 0 to 70°C, which is satisfactory for most commercial and industrial applications. IC's which pass this test are then usually tested in the range -55°C to $+125$°C, and circuits passing tests in this range are then suitable for most military and space applications. This provides two grades of IC.

11.5 Linear integrated circuits

Linear or analogue IC's can have an output of any value (within device limits) proportional to the input, and the output changes in a smooth manner as the input is changed. This contrasts with digital IC's, which are either on or off.

Linear IC's are used to amplify current, voltage, or power.

11.6 General-purpose linear IC's

The circuits which perform the basic amplifier functions are known as operational amplifiers, or just op-amps, and are high-gain direct-coupled wideband amplifiers. Op-amps are assumed to have the following 'ideal' properties:

a) infinite gain,
b) infinite bandwidth,
c) infinite input impedance,
d) zero output impedance,
e) constant phase shift between input and output.

In practice the op-amp does not have these ideal characteristics, but the actual performance of the device is modified by the use of external components, normally using negative feedback, so that this is not a problem.

Modern linear IC's have extremely complex internal circuitry. The internal circuitry for a relatively old and 'simple' IC – the 741 op-amp – is shown in the set of characteristics for the μA741 op-amp at the end of the chapter.

Access to the internal circuitry is not possible – the device comes in standard packages, shown in fig. 11.1, the most common being the eight-pin DIL (dual in-line) package. The op-amp symbol is shown in fig. 11.5. Note that the convention is to omit power-supply pins on the symbol, since it is taken for granted that they must be available.

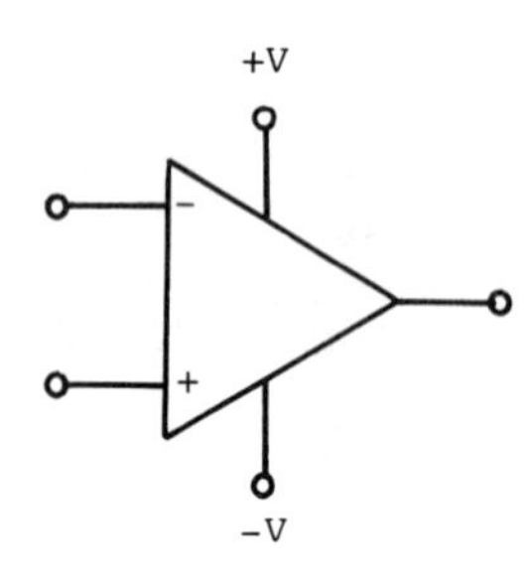

Fig. 11.5 Operational-amplifier circuit

11.7 Performance characteristics of linear IC's

The range of modern linear IC's and the functions they perform is enormous. The identification from part numbers is also quite complex. Tables 11.1 to 11.4 explain some of the codes used.

Table 11.1 IC part-number description

Part number	Cross ref part no.	Product family	Product description
NE5534N		ANA	Low-noise op-amp
μA741C	LM741CJ	ANA	General-purpose op-amp

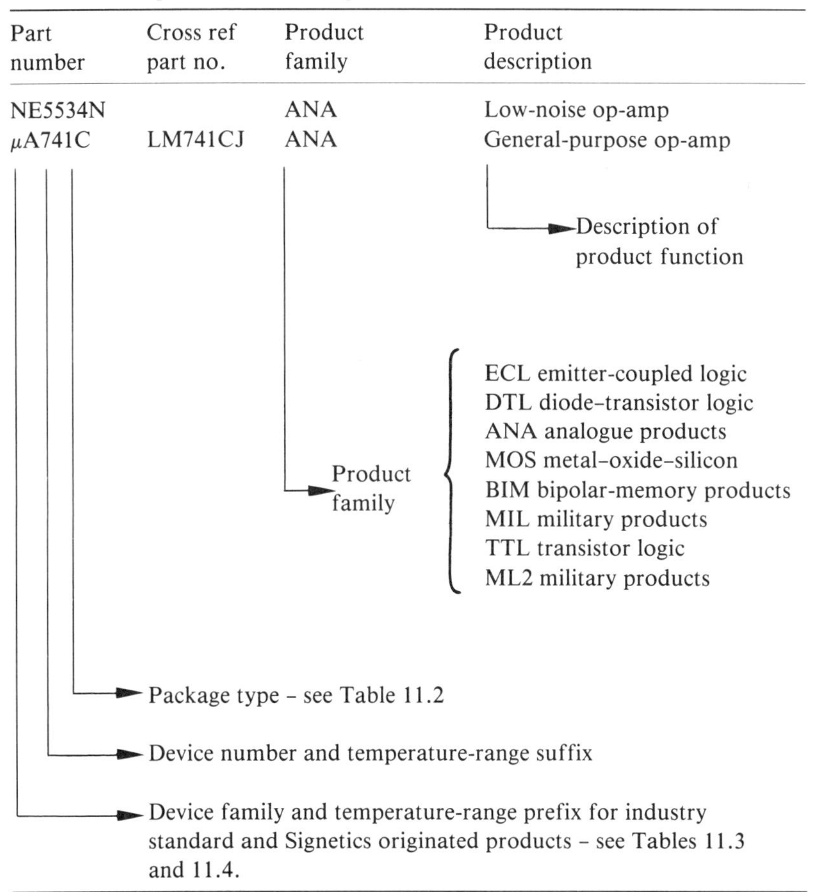

The performance characteristics of each device are provided by the manufacturers in comprehensive data books. To give some idea of what is involved, extracts from data sheets are provided at the end of the chapter for some typical commercially available devices listed below:

a) general-purpose operational amplifier μA741,
b) AM receiver circuit TDA1072,
c) comparator/differential-amplifier LM111,
d) r.f./i.f. amplifier CA3089,
e) wideband amplifier NE/SE5539.

Table 11.2 IC package descriptions

Suffix		Package
Old	New	description
A, AA	N	14-lead plastic DIL
A	N-14	14-lead plastic DIL (selected analogue products only)
B, BA	N	16-lead plastic DIL
—	D	Microminiature package (SO)
F	F	14, 16, 18, 22, and 24-lead ceramic (Cerdip) DIL
I, IK	I	14, 16, 18, 22, 28, and 4-lead ceramic DIL
K	H	10-lead TO-100
L	H	10-lead high-profile TO-100 can
NA, NX	N	24-lead plastic DIL
Q, R	Q	10, 14, 16, and 24-lead ceramic flat
T, TA	H	8-lead TO-99
U	U	SIL plastic power
V	N	8-lead plastic DIL
W, WJ	W	10, 14, 16, and 24 lead ceramic (Cerpac) flat
XA	N	18-lead plastic DIL
XC	N	20-lead plastic DIL
XC	N	22-lead plastic DIL
XL, XF	N	28-lead plastic DIL

Table 11.3 IC family prefix

Prefix	Device family
CA	Linear industry standard
DS	Linear industry standard
JB	Mil rel – JAN qualified* – old designator
JM	Mil rel – JAN qualified* – new designator
LH	Linear industry standard
LM	Linear industry standard
M	Mil rel – JAN processed
MC	Linear industry standard
SD	Linear DMOS
μA	Linear industry standard
ULN	Linear industry standard

* 'JAN qualified' indicates that the device has been manufactured to the USA government military specifications and has been monitored by the USA Defense Electronic Supply Center during manufacture.

Table 11.4 IC device temperature

Prefix	Device temperature range
N-	0° to +70°C
S-	−55° to +125°C
NE-	0° to +70°C
SE-	−55° to +125°C
SA	−40° to +85°C
SU	−25° to +85°C

11.8 The basic amplifier

Two of the simplest applications for op-amps are as inverting and non-inverting amplifiers. The basic arrangement for the inverting amplifier is shown in fig. 11.6.

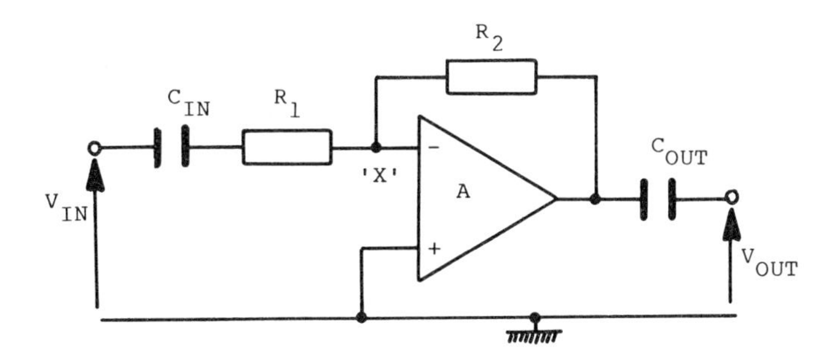

Fig. 11.6 Inverting amplifier

A resistor R_1 is connected between the input terminal and the inverting input. (Note: the ' − ' and ' + ' on the amplifier symbol simply indicate the inverting and non-inverting inputs respectively.) Feedback from output to input is via resistor R_2. The non-inverting input is grounded. With typical gain figures (in excess of 100 000), an output voltage of 10 V is achieved with an input of $V_{IN} = V_{OUT}/A = 10/100\,000 = 100\ \mu V$. This is so small that point 'X' in fig. 11.6 is virtually at earth potential – this is known as the 'virtual-earth' concept.

The very high gain is controlled by the negative feedback. The input voltage V_{IN} appears across R_1 and the input current is V_{IN}/R_1, since point 'X' is at ground potential (this may be at earth potential, but not necessarily – e.g. in portable equipment).

Since the input resistance is very high, negligible current flows into the op-amp itself; therefore all the current flows in R_2. The current in R_2 is also found from V_{OUT}/R_2.

$$\therefore \quad \frac{V_{IN}}{R_1} = \frac{-V_{OUT}}{R_2}$$

$$\therefore \quad A' = \frac{V_{OUT}}{V_{IN}} = \frac{-R_2}{R_1} \qquad 11.1$$

From equation 11.1 we can see that the gain depends only on the value of the resistors R_1 and R_2.

Since point 'X' is at ground potential, the circuit input resistance is effectively R_1; the circuit is using current series feedback.

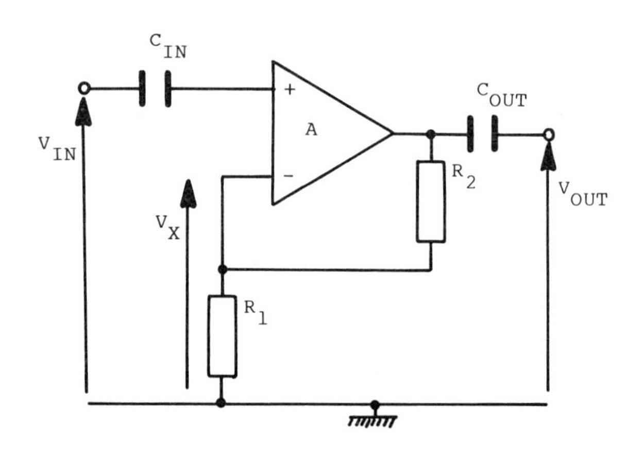

Fig. 11.7 Non-inverting amplifier

Figure 11.7 shows the basic non-inverting amplifier circuit. The voltage between the amplifier terminals is $V_{IN} - V_X$.

$$\therefore \quad V_{OUT} = A(V_{IN} - V_X) \qquad 11.2$$

and $\quad V_X = \dfrac{V_{OUT}R_1}{R_1 + R_2}$

Substituting for V_X in equation 11.2, we get

$$V_{OUT} = A\left(V_{IN} - \frac{V_{OUT}R_1}{R_1 + R_2}\right)$$

$$\therefore \quad V_{OUT}\left(1 + \frac{AR_1}{R_1 + R_2}\right) = AV_{IN}$$

$$\therefore \quad A' = \frac{V_{OUT}}{V_{IN}} = \frac{A}{1 + A R_1/(R_1 + R_2)}$$

$$= \frac{1}{1/A + R_1/(R_1 + R_2)}$$

If $A \gg 1$ then $1/A \rightarrow 0$

$$\therefore \quad A' = \frac{R_1 + R_2}{R_1}$$

This is voltage series feedback. The gain still depends only on the value of external resistors. The input resistance is very high, since it is the input resistance of the op-amp itself. The output resistance of the circuit is R_1 in series with the parallel combination of R_2 and the output resistance of the op-amp itself (75 Ω in the case of the 741). In most cases this means that it will effectively be R_1.

Problems

11.1 The inverting op-amp of fig. 11.6 has R_1 = 15 kΩ and R_2 = 120 kΩ. Calculate the gain of the circuit.

11.2 The non-inverting amplifier of fig. 11.7 has R_1 = 2.2 kΩ and R_2 = 18 kΩ. Calculate the gain of the circuit.

11.3 The amplifier of fig. 11.6 is required to have an input resistance of 100 kΩ and a gain of 200. Calculate

a) the required values for R_1 and R_2,

b) the feedback fraction β if the gain of the op-amp is 200 000.

11.4 The amplifier of fig. 11.7 must have an output resistance of 2 kΩ and a gain of 60. Calculate suitable values for R_1 and R_2.

11.5 Why are integrated circuits generally low-voltage devices?

11.6 For the 741 op-amp, compare and comment on any differences between the following 'ideal' and actual properties:

a) gain,

b) bandwidth,

c) input impedance,

d) output impedance.

11.7 For the SE5539 device, use the information provided in the data sheets to determine:

a) the supply pins,

b) the supply voltage,

c) the temperature range over which the device can be expected to operate satisfactorily,

d) the type of packaging,

e) the bandwidth for small-signal operation.

11.8 Explain briefly what is meant by the term 'virtual earth' in relation to op-amps.

11.9 List the advantages of using op-amps instead of the equivalent discrete circuit arrangements.

11.10 What is the practical difference between the inverting and non-inverting inputs of an op-amp?

11.11 How are capacitors fabricated on integrated circuits?

11.12 Give three applications where integrated circuits would not be suitable.

GENERAL PURPOSE OPERATIONAL AMPLIFIER μA741/ μA741C/SA741C

DESCRIPTION

The μA741 is a high performance operational amplifier with high open loop gain, internal compensation, high common mode range and exceptional temperature stability. The μA741 is short-circuit protected and allows for nulling of offset voltage.

FEATURES

- Internal frequency compensation
- Short circuit protection
- Excellent temperature stability
- High input voltage range

PIN CONFIGURATION

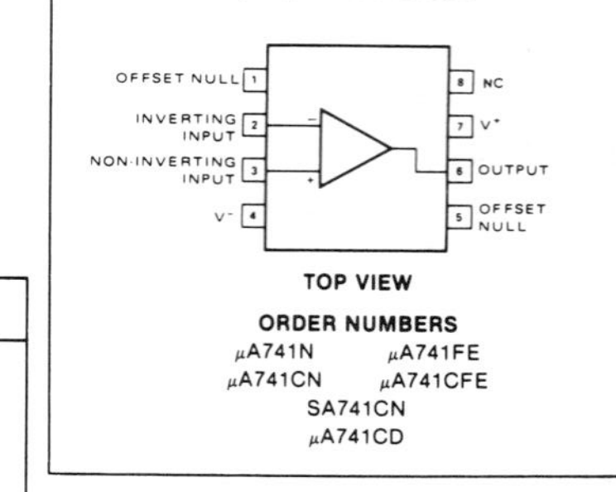

ABSOLUTE MAXIMUM RATINGS

PARAMETER	RATING	UNIT
Supply voltage		
μA741C	±18	V
μA741	±22	V
Internal power dissipation		
N package	500	mW
FE package	1000	mW
Differential input voltage	±30	V
Input voltage[1]	±15	V
Output short-circuit duration	Continuous	
Operating temperature range		
μA741C	0 to +70	°C
SA741C	-40 to +85	°C
μA741	-55 to +125	°C
Storage temperature range	-65 to +150	°C
Lead temperature (soldering 60sec)	300	°C

NOTE

1. For supply voltages less than ±15V, the absolute maximum input voltage is equal to the supply voltage.

μA741 (continued)

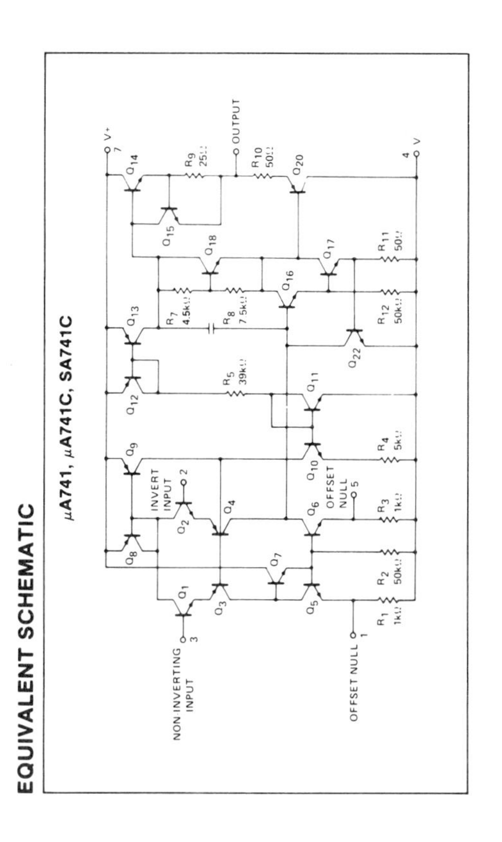

DC ELECTRICAL CHARACTERISTICS T_A = 25°C, V_S = ±15V, unless otherwise specified.

	PARAMETER	TEST CONDITIONS	μA741			μA741C			UNIT
			Min	Typ	Max	Min	Typ	Max	
V_{OS}	Offset voltage	R_S = 10kΩ		1.0	5.0		2.0	6.0	mV
		R_S = 10kΩ, over temp.		1.0	6.0			7.5	mV
I_{OS}	Offset current			20	200		20	200	nA
		Over temp.						300	nA
		T_A = +125°C		7.0	200				nA
		T_A = -55°C		20	500				nA
I_{BIAS}	Input bias current			80	500		80	500	nA
		Over temp.						800	nA
		T_A = +125°C		30	500				nA
		T_A = -55°C		300	1500				nA
V_{OUT}	Output voltage swing	R_L = 10kΩ	±12	±14		±12	±14		V
		R_L = 2kΩ, over temp.	±10	±13		±10	±13		V
A_{VOL}	Large signal voltage gain	R_L = 2kΩ, V_O = ±10V	50	200		20	200		V/mV
		R_L = 2kΩ, V_O = ±10V, over temp.	25			15			V/mV
	Offset voltage adjustment range			±30			±30		mV
PSRR	Supply voltage rejection ratio	R_S ≤ 10kΩ					10	150	μV/V
		R_S ≤ 10kΩ, over temp.		10	150				μV/V
CMRR	Common mode rejection ratio								dB
		Over temp.	70	90					dB
I_{CC}	Supply current			1.4	2.8		1.4	2.8	mA
		T_A = +125°C		1.5	2.5				mA
		T_A = -55°C		2.0	3.3				mA
V_{IN}	Input voltage range	(μA741, over temp.)	±12	±13		±12	±13		V
R_{IN}	Input resistance		0.3	2.0		0.3	2.0		MΩ
P_d	Power consumption			50	85		50	85	mW
		T_A = +125°C		45	75				mW
		T_A = -55°C		45	100				mW
R_{OUT}	Output resistance			75			75		Ω
I_{SC}	Output short-circuit current			25			25		mA

DC ELECTRICAL CHARACTERISTICS (Cont'd) $T_A = 25°C$, $V_S = \pm15V$, unless otherwise specified.

	PARAMETER	TEST CONDITIONS	SA741C			UNIT
			Min	Typ	Max	
V_{OS}	Offset voltage	$R_S = 10k\Omega$		2.0	6.0	mV
		$R_S = 10k\Omega$, over temp.			7.5	mV
I_{OS}	Offset current			20	200	nA
		Over temp.			500	nA
I_{BIAS}	Input bias current			80	500	nA
		Over temp.			1500	nA
V_{OUT}	Output voltage swing	$R_L = 10k\Omega$	±12	±14		V
		$R_L = 2k\Omega$, over temp.	±10	±13		V
A_{VOL}	Large signal voltage gain	$R_L = 2k\Omega$, $V_O = \pm10V$	20	200		V/mV
		$R_L = 2k\Omega$, $V_O = \pm10V$, over temp.	15			V/mV
	Offset voltage adjustment range			±30		mV
PSRR	Supply voltage rejection ratio	$R_S \leq 10k\Omega$		10	150	μV/V
CMRR	Common mode rejection ratio					dB
I_{CC}	Supply current			1.4	2.8	mA
V_{IN}	Input voltage range	(μA741, over temp.)	±12	±13		V
	Input resistance		0.3	2.0		MΩ
P_d	Power consumption			50	85	mW
R_{OUT}	Output resistance			75		Ω
I_{SC}	Output short-circuit current			25		mA

AC ELECTRICAL CHARACTERISTICS

$T_A = 25°C$, $V_S = ±15V$, unless otherwise specified.

PARAMETER	TEST CONDITIONS	μA741, μA741C			UNIT
		Min	Typ	Max	
Parallel input resistance	Open loop, f = 20Hz				MΩ
Parallel input capacitance	Open loop, f = 20Hz		1.4		pF
Unity gain crossover frequency	Open loop		1.0		MHz
Transient response unity gain	V_{IN} = 20mV, R_L = 2kΩ, C_L ≤ 100pf				
Rise time			0.3		μs
Overshoot			5.0		%
Slew rate	C ≤ 100pf, R_L ≥ 2k, V_{IN} = ±10V		0.5		V/μs

TYPICAL PERFORMANCE CHARACTERISTICS

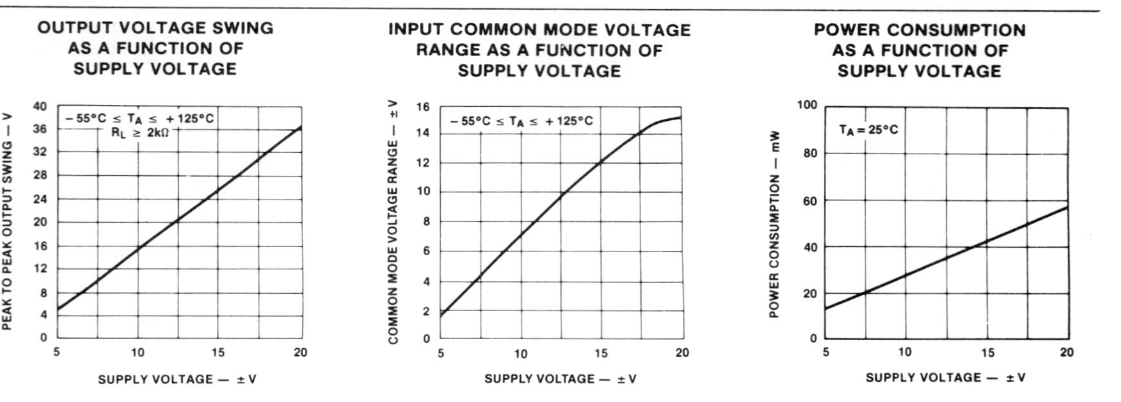

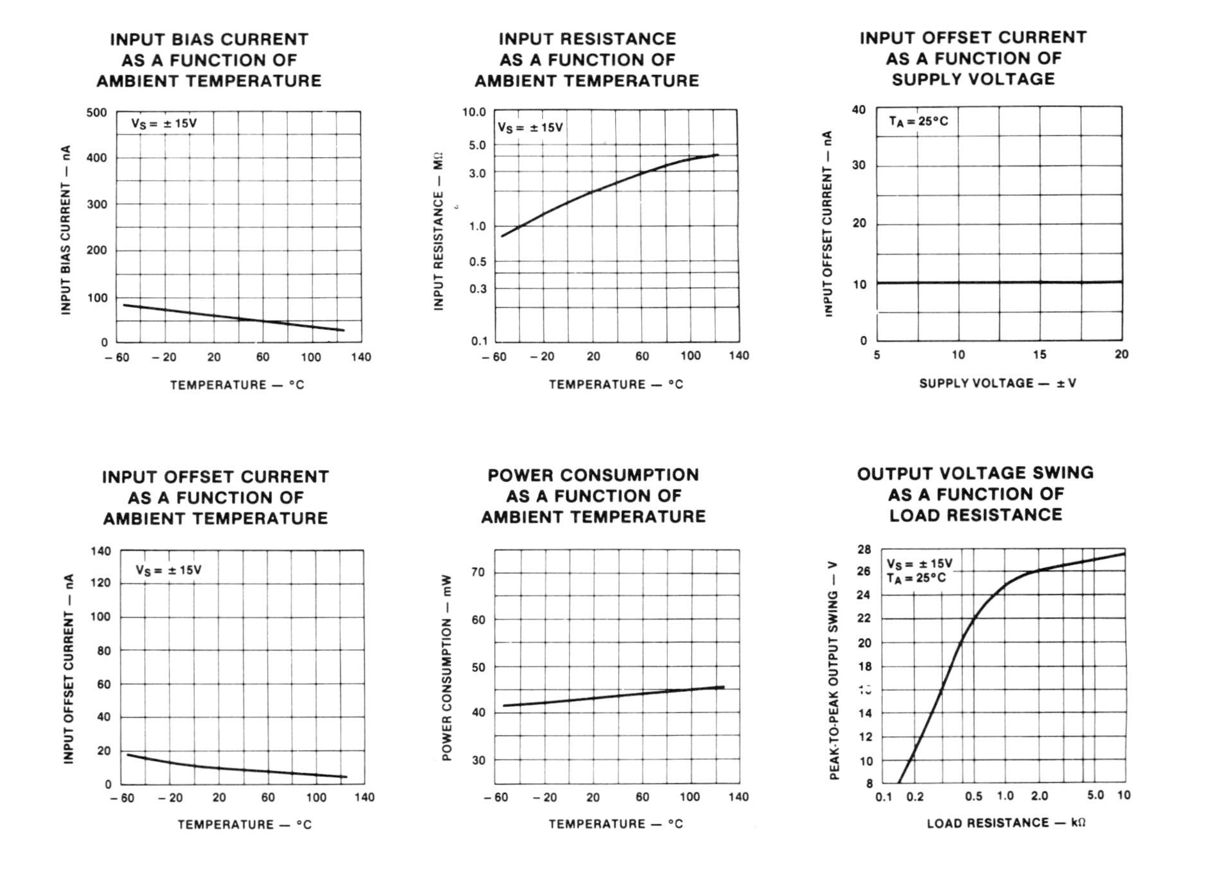
INPUT BIAS CURRENT AS A FUNCTION OF AMBIENT TEMPERATURE
$V_S = \pm 15V$
INPUT BIAS CURRENT — nA
500 400 300 200 100 0
−60 −20 20 60 100 140
TEMPERATURE — °C
INPUT RESISTANCE AS A FUNCTION OF AMBIENT TEMPERATURE
$V_S = \pm 15V$
INPUT RESISTANCE — MΩ
10.0 5.0 3.0 1.0 0.5 0.3 0.1
−60 −20 20 60 100 140
TEMPERATURE — °C
INPUT OFFSET CURRENT AS A FUNCTION OF SUPPLY VOLTAGE
$T_A = 25°C$
INPUT OFFSET CURRENT — nA
40 30 20 10 0
5 10 15 20
SUPPLY VOLTAGE — ±V
INPUT OFFSET CURRENT AS A FUNCTION OF AMBIENT TEMPERATURE
$V_S = \pm 15V$
INPUT OFFSET CURRENT — nA
140 120 100 80 60 40 20 0
−60 −20 20 60 100 140
TEMPERATURE — °C
POWER CONSUMPTION AS A FUNCTION OF AMBIENT TEMPERATURE
POWER CONSUMPTION — mW
70 60 50 40 30
−60 −20 20 60 100 140
TEMPERATURE — °C
OUTPUT VOLTAGE SWING AS A FUNCTION OF LOAD RESISTANCE
$V_S = \pm 15V$
$T_A = 25°C$
PEAK-TO-PEAK OUTPUT SWING — V
28 26 24 22 20 18 14 12 10 8
0.1 0.2 0.5 1.0 2.0 5.0 10
LOAD RESISTANCE — kΩ

μA741 (continued)

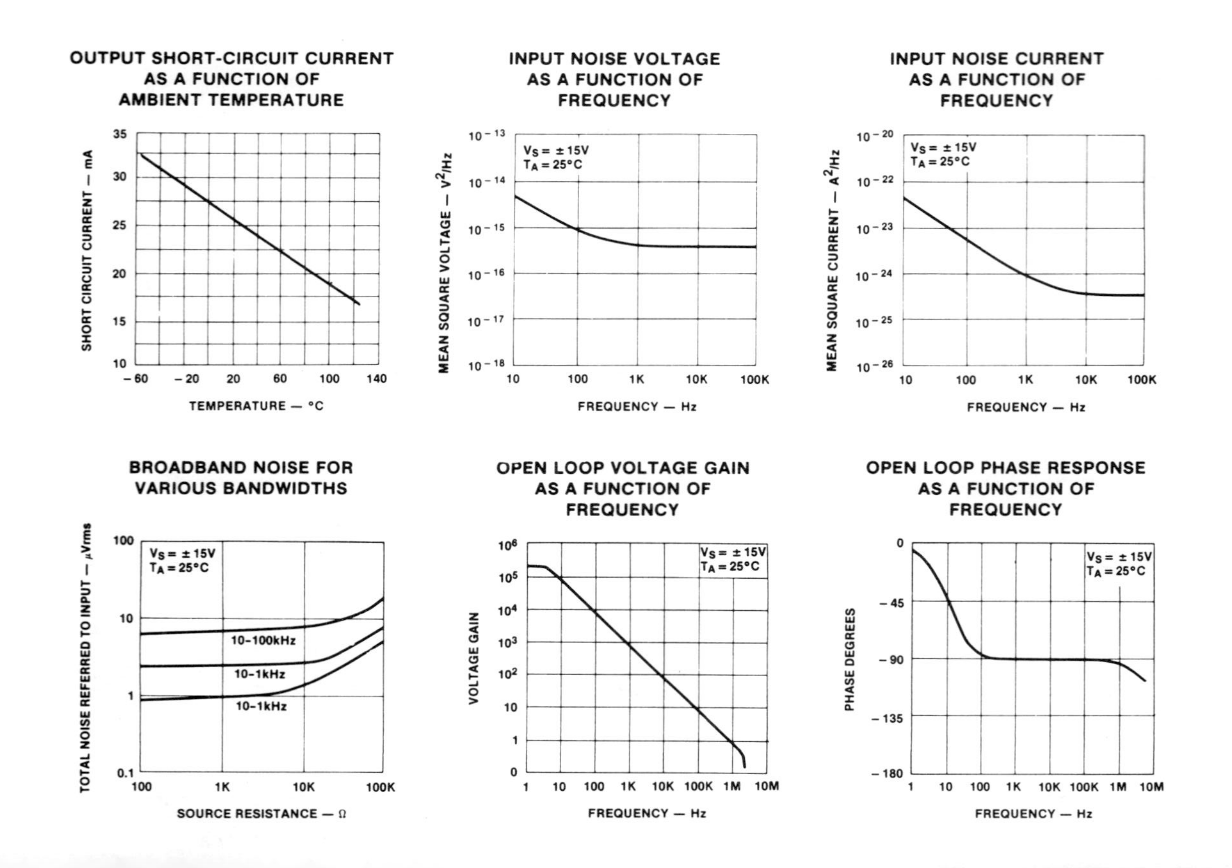

OUTPUT VOLTAGE SWING
AS A FUNCTION OF
FREQUENCY

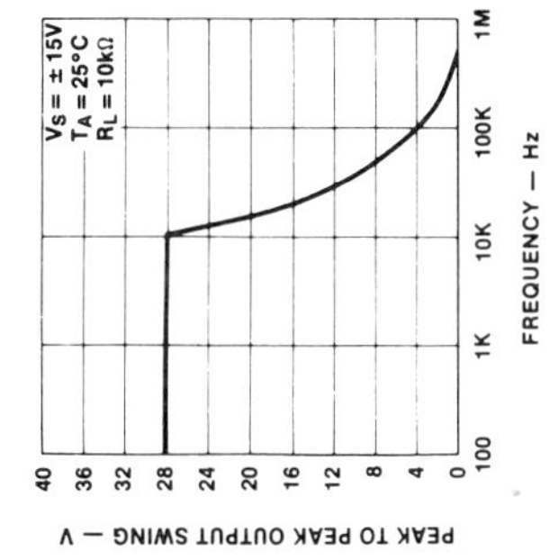
$V_S = \pm 15V$
$T_A = 25°C$
$R_L = 10k\Omega$
PEAK TO PEAK OUTPUT SWING — V
FREQUENCY — Hz
40
36
32
28
24
20
16
12
8
4
0
100
1K
10K
100K
1M

COMMON MODE REJECTION
RATIO AS A FUNCTION OF
FREQUENCY

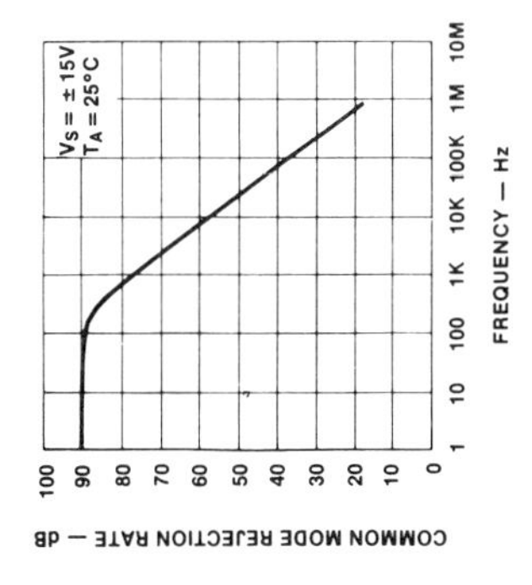
$V_S = \pm 15V$
$T_A = 25°C$
COMMON MODE REJECTION RATE — dB
FREQUENCY — Hz
100
90
80
70
60
50
40
30
20
10
0
1
10
100
1K
10K
100K
1M
10M

TRANSIENT RESPONSE

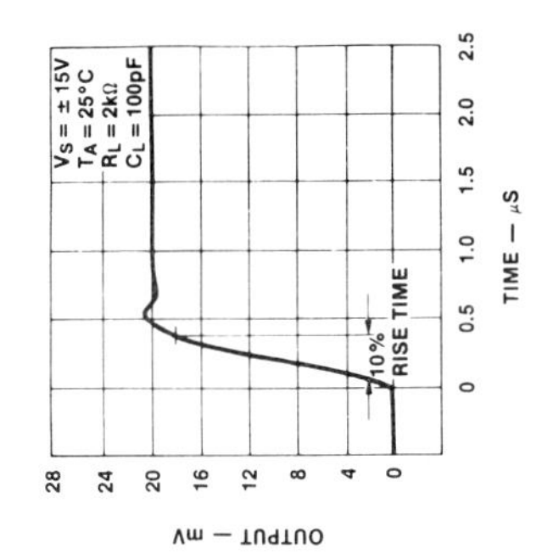
$V_S = \pm 15V$
$T_A = 25°C$
$R_L = 2k\Omega$
$C_L = 100pF$
10%
RISE TIME
OUTPUT — mV
TIME — μS
28
24
20
16
12
8
4
0
0
0.5
1.0
1.5
2.0
2.5

POWER BANDWIDTH
(Large Signal Swing vs Frequency)

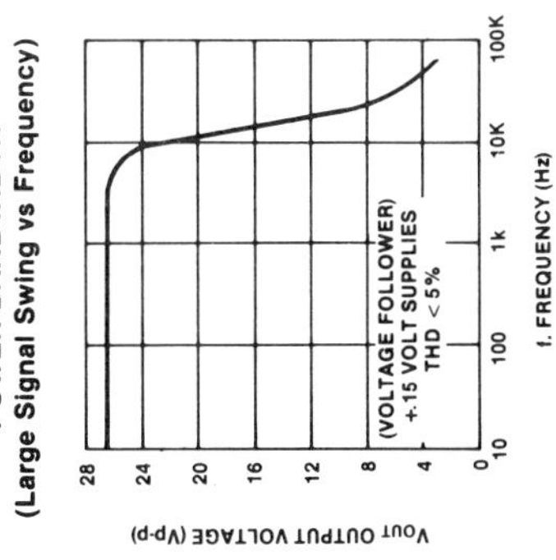
(VOLTAGE FOLLOWER)
+.15 VOLT SUPPLIES
THD <5%
V_{OUT} OUTPUT VOLTAGE (Vp-p)
f. FREQUENCY (Hz)
28
24
20
16
12
8
4
0
10
100
1k
10k
100K

AM RECEIVER CIRCUIT

The TDA1072 is a monolithic integrated AM receiver circuit provided with the following functions:

- controlled h.f. preamplifier
- multiplicative balanced mixer
- separate oscillator with amplitude control
- i.f. amplifier with gain control
- balanced full-wave detector
- a.f. preamplifier
- internal a.g.c. voltage
- amplifier for field-strength indication
- electronic stand-by on/off switch

QUICK REFERENCE DATA

Parameter	Symbol		Value	Unit
Supply voltage (pin 13)	V_P	typ.	15	V
Supply current	I_P	typ.	22	mA
H.F. input voltage				
S + N/N = 6 dB	V_i	typ.	2,2	μV
S + N/N = 26 dB	V_i	typ.	30	μV
H.F. input voltage; d_{tot} = 3%; m = 80%	V_i	typ.	650	mV
A.F. output voltage; V_i = 2 mV	V_o	typ.	340	mV
Total distortion	d_{tot}	typ.	0,5	%
Input voltage range for ΔV_o = 6 dB	ΔV_i	typ.	91	dB
Oscillator frequency range	f_{osc}		0,6 to 31	MHz
Oscillator voltage amplitude	V_{osc}	typ.	140	mV
Field-strength indication range	ΔV_i	typ.	100	dB
Supply voltage range	V_P		7,5 to 18	V
Ambient temperature range	T_{amb}		−30 to + 80	°C

PACKAGE OUTLINE

16-lead DIL; plastic (SOT-38).

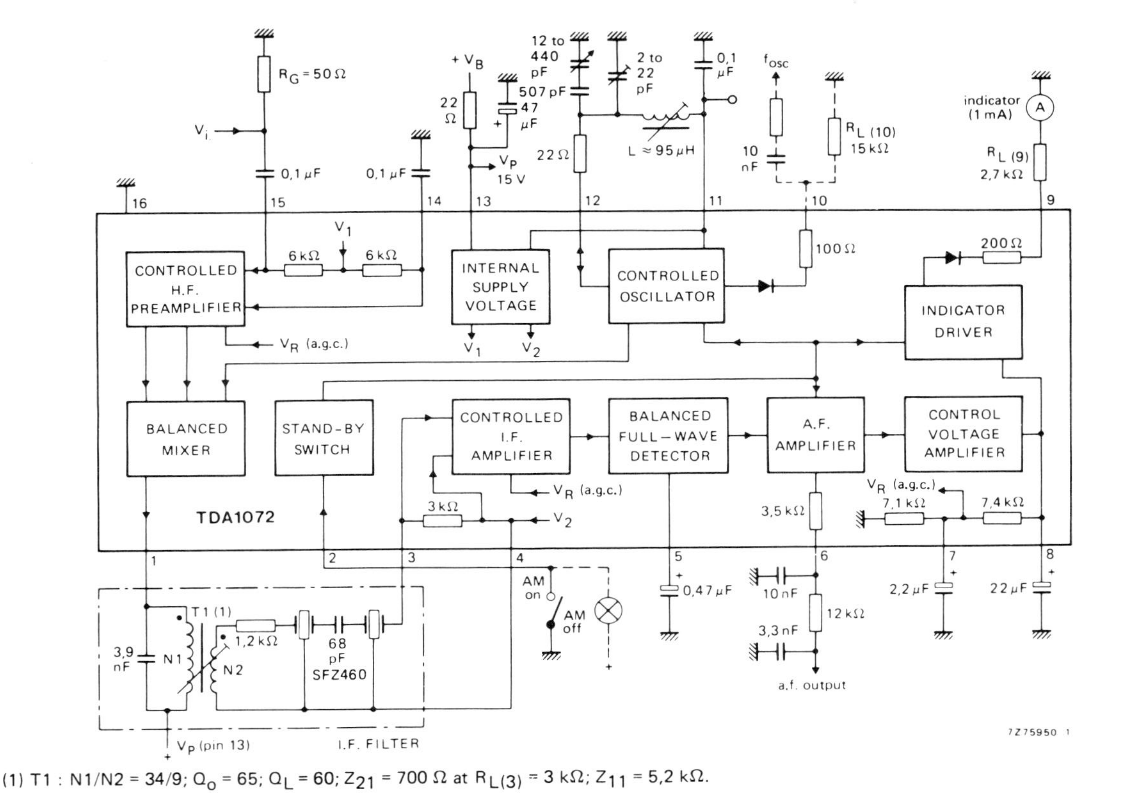

(1) T1 : N1/N2 = 34/9; Q_o = 65; Q_L = 60; Z_{21} = 700 Ω at $R_{L(3)}$ = 3 kΩ; Z_{11} = 5,2 kΩ.

Fig. 1 Block diagram with external components; used as test circuit.

RATINGS

Limiting values in accordance with the Absolute Maximum System (IEC 134)

Supply voltage (pin 13)	$V_P = V_{13\text{-}16}$	max.	23 V
Voltage on pin 2	$V_{2\text{-}16}$		0 to 23 V

H.F. inputs

Voltages between:			
pins 14 and 15	$\pm V_{14\text{-}15}$	max.	12 V
pins 14 and 16	$V_{14\text{-}16}$	max.	V_P V
pins 15 and 16	$V_{15\text{-}16}$	max.	V_P V
Or currents:			
pin 14	$\pm I_{14}$	max.	10 mA
pin 15	$\pm I_{15}$	max.	10 mA
Storage temperature range	T_{stg}		−55 to + 150 °C
Operating ambient temperature range	T_{amb}		−30 to + 80 °C

CHARACTERISTICS

V_P = 15 V; T_{amb} = 25 °C; f_i = 1 MHz (h.f.), R_G = 50 Ω; f_m = 0,4 kHz; m = 30%;
i.f. frequency = 460 kHz; unless otherwise specified

Supply voltage range (pin 13)	V_P		7,5 to 18 V
Supply current; without load ($I_{L(11)}$ = 0)	I_P	typ.	22 mA
			15 to 30 mA

H.F. preamplifier and mixer

D.C. input voltages	$V_{14\text{-}16}$; $V_{15\text{-}16}$	typ.	2,75 ($4V_{BE}$) V
Input impedance			
$V_i < 300\ \mu V$	$Z_{i(14\text{-}16)}$; $Z_{i(15\text{-}16)}$	typ.	6 kΩ
		typ.	6 pF
$V_i > 10$ mV	$Z_{i(14\text{-}16)}$; $Z_{i(15\text{-}16)}$	typ.	9 kΩ
		typ.	2,5 pF
Output impedance	$Z_{o(1\text{-}16)}$	>	200 kΩ
		typ.	4 pF
Maximum conversion conductance	S_M	typ.	5,5 mA/V*
Maximum i.f. output voltage (peak-to-peak value)	$V_{o(1)(p\text{-}p)}$	typ.	2,8 V
Output current capability	$I_{o(1)}$	typ.	1 mA
Control range of preamplifier	ΔS_M	typ.	30 dB
Maximum h.f. input voltage (peak-to-peak value)	$V_{i(14\text{-}15)(p\text{-}p)}$	typ.	2,8 V

* S_M is defined as $I_{o(1)}/V_i$.

CHARACTERISTICS (continued)

Parameter	Symbol		Value	Unit
Oscillator				
Frequency range	$f_{osc(12)}$		0,6 to 31	MHz
Oscillator impedance range	$Z_{L(12)}$		1 to 200	kΩ
Controlled oscillator amplitude	$V_{osc(12)}$	typ.	140	mV
		<	200	mV
D.C. output voltage ($I_{L(11)} = 0$)	$V_{11\text{-}16}$	typ.	$V_P - 1,3$	V
Output load current range	$-I_{L(11)}$		0 to 15	mA
Output resistance; $I_{L(11)} = 5 \pm 0,5$ mA	$R_{o(11)}$	typ.	7	Ω
Oscillator frequency output (pin 10)				
Output voltage (peak-to-peak value) $R_{10\text{-}16} = 15$ kΩ ($R_{L(10)}$)	$V_{o(10)(p\text{-}p)}$	typ.	200	mV
Output resistance	$R_{o(10)}$	typ.	150	Ω
Allowable output current (peak value)	$I_{o(10)M}$	<	2	mA
I.F. amplifier and a.f. stage				
D.C. input voltages	$V_{3\text{-}16}$; $V_{4\text{-}16}$	typ.	2	V
Input impedance	$Z_{i(3)}$	typ.	3	kΩ
			2,4 to 3,9	kΩ
		typ.	4	pF
Max. i.f. input voltage; m = 80%; $d_{tot} = 3\%$	$V_{i(3)}$	typ.	75	mV
Control range; $V_o = -6$ dB	ΔV_i	typ.	62	dB
A.F. output voltage; $V_{i(3)} = 2$ mV; without load	$V_{o(6)}$	typ.	350	mV
A.F. output resistance	$R_{o(6)}$	typ.	3,5	kΩ
Field-strength indication				
D.C. indicator voltage $V_i = 0$; $R_{L(9)} = 2,7$ kΩ	$V_{9\text{-}16}$	typ.	0	mV
		<	140	mV
$V_i = 500$ mV; $R_{L(9)} = 2,7$ kΩ	$V_{9\text{-}16}$	typ.	2,8	V
			2,5 to 3,1	V
Output current capability	$-I_9$	>	1,2	mA
Output resistance; $-I_9 = 0,5$ mA	$R_{o(9)}$	typ.	250	Ω
Leakage voltage at the output; $\pm I_9 \leqslant 1$ µA; at AM switch off ($V_{2\text{-}16} \geqslant 3,5$ V)	$V_{9\text{-}16}$	typ.	6	V

Stand-by switch

Switching voltage	$V_{2\text{-}16}$	typ.	2,6 V
Required control voltage*			
AM on	$V_{2\text{-}16}$	<	2 V
AM off	$V_{2\text{-}16}$	>	3,5 V**
Input current			
AM on; switching current	$-I_2$	<	100 μA
AM off; leakage current ($V_{2\text{-}16} = V_{3\text{-}16}$)	$\pm I_2$	<	1 μA

APPLICATION INFORMATION

V_P = 15 V; T_{amb} = 25 °C; measured in Fig. 1; f_i = 1 MHz (h.f.); f_m = 0,4 kHz; m = 30%; unless otherwise specified

H.F. input voltage			
S + N/N = 6 dB	V_i	typ.	2,2 μV
S + N/N = 10 dB	V_i	typ.	3,5 μV
S + N/N = 26 dB	V_i	typ.	30 μV
S + N/N = 46 dB	V_i	typ.	550 μV
H.F. input voltage for a.g.c. operation	V_i	typ.	14 μV
Control range for ΔV_o = 6 dB			
reference value V_i = 500 mV	ΔV_i	typ.	91 dB
Maximum h.f. input voltage			
d_{tot} = 3%; m = 80%	V_i	typ.	0,65 V
d_{tot} = 3%; m = 30%	V_i	typ.	0,9 V
d_{tot} = 10%; m = 30%	V_i	typ.	1,3 V
A.F. output voltage; V_i = 2 mV	V_o	typ.	340 mV
Change of a.f. output voltage; V_i = 2 mV	ΔV_o	typ.	± 2 dB
H.F. input voltage; V_o = 60 mV	V_i	typ.	4 μV
Total distortion of a.f. output voltage			
V_i = 2 mV; m = 80%	d_{tot}	typ.	0,5 %
V_i = 500 mV; m = 80%	d_{tot}	typ. <	1,8 % 3 %
Signal plus noise-to-noise ratio of a.f. output voltage			
V_i = 2 mV	S + N/N	typ.	50 dB
I.F. bandwidth (−3 dB)	B	typ.	4,6 kHz
I.F. selectivity			
Δf = ± 9 kHz	$S_{(9)}$	typ.	30 dB
Δf = ± 36 kHz	$S_{(36)}$	typ.	60 dB

* At allowable ambient temperature range and supply voltage range.
** Also achieved at open input.

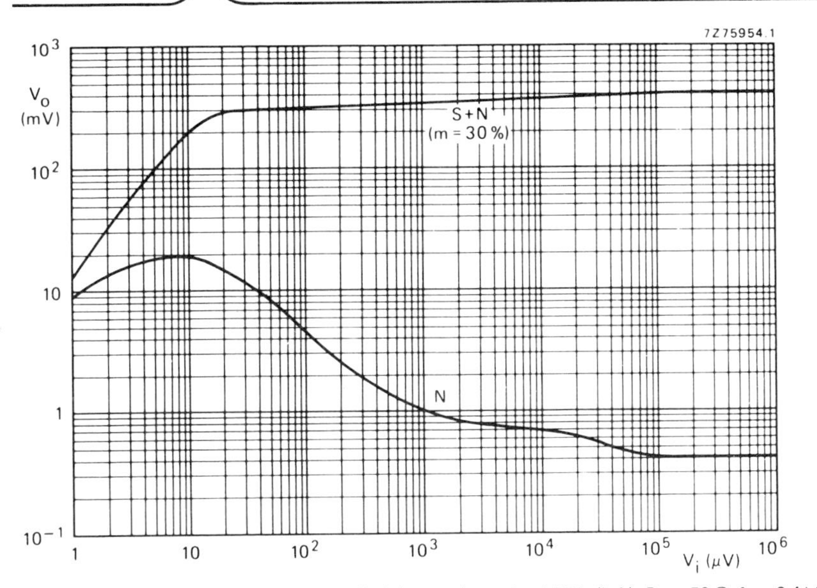

Fig. 2 A.F. output voltage as a function of h.f. input voltage; f_i = 1 MHz (h.f.); R_G = 50 Ω; f_m = 0,4 kHz.

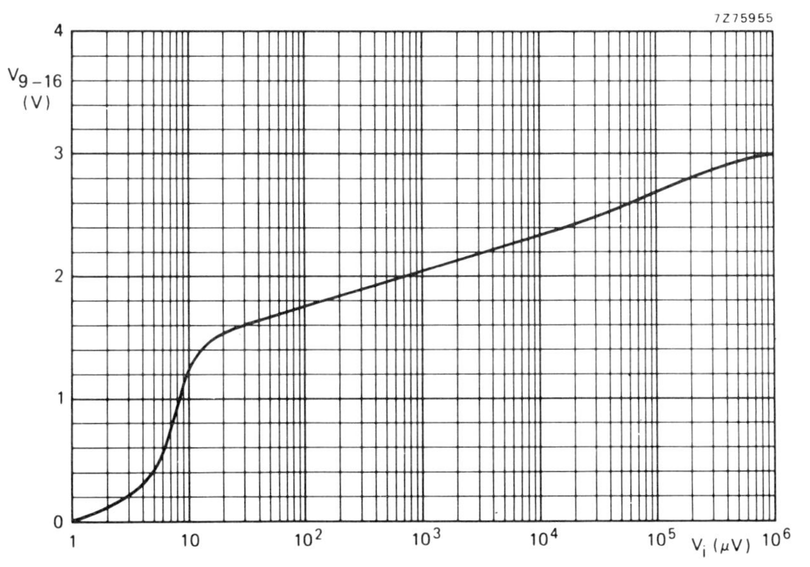

Fig. 3 Indication voltage as a function of h.f. input voltage; R_{9-16} = 2,7 kΩ.

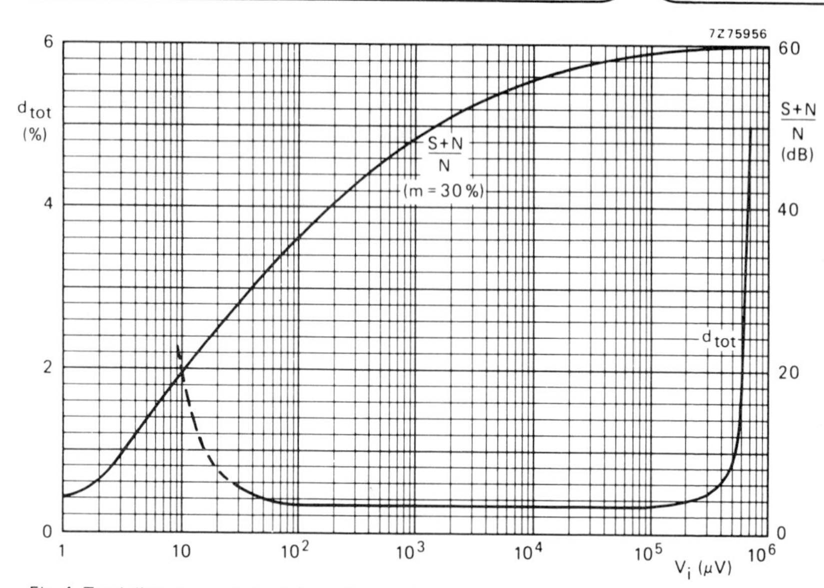

Fig. 4 Total distortion and signal plus noise-to-noise ratio as a function of h.f. input voltage; for d_{tot} : f_m = 0,4 kHz; m = 80%.

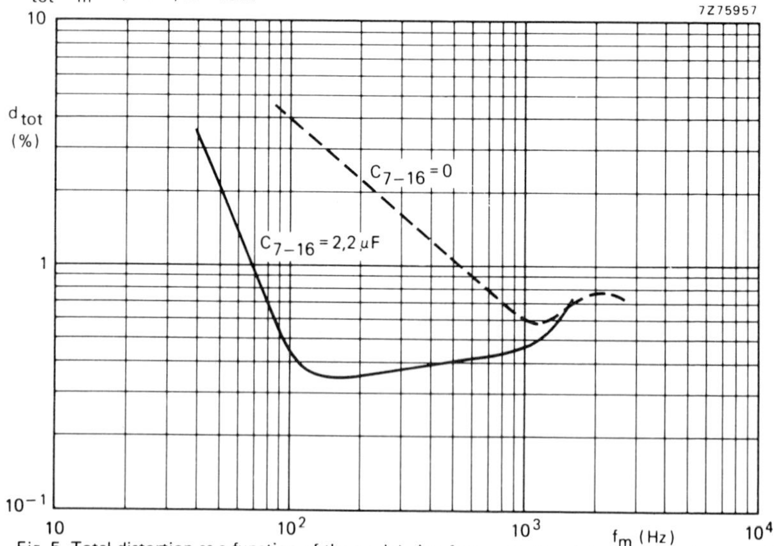

Fig. 5 Total distortion as a function of the modulation frequency; V_i = 10 mV; f_i = 1 MHz; m = 80%. $C_{8\text{-}16}$ = 22 μF.

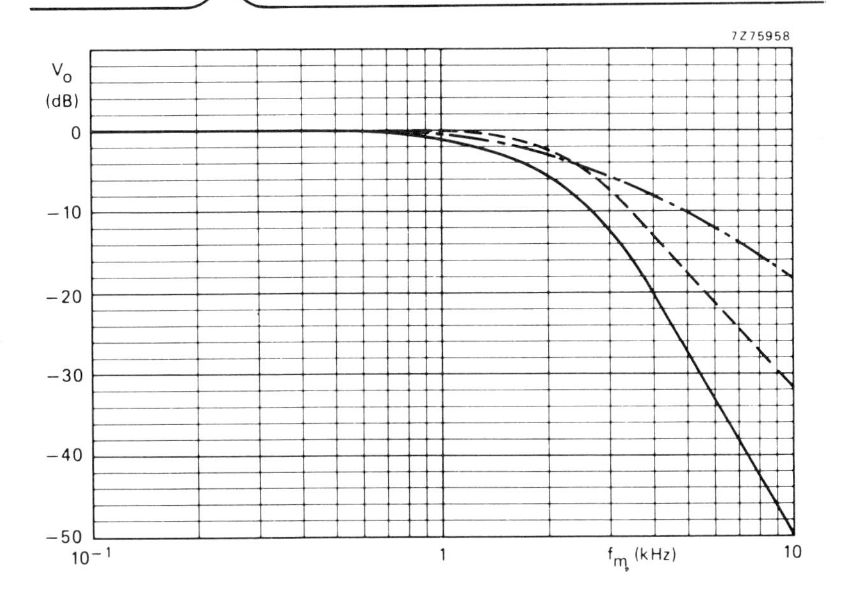

Fig. 6 Frequency responses (wobbled) for various conditions:
——— with a.f. and i.f. filter
– – – with i.f. filter
– · – with a.f. filter

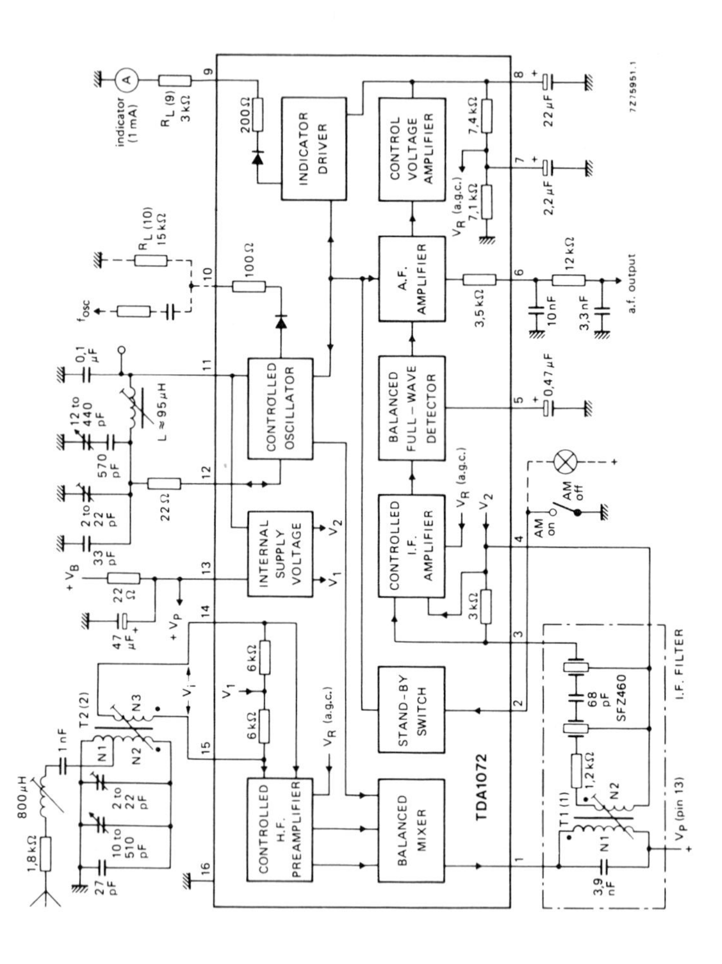

(1) T1 : N1/N2 = 34/9; Q_o = 65; Q_L = 60; Z_{21} = 700 Ω at $R_{L(3)}$ = 3 kΩ; Z_{11} = 5,2 kΩ.
(2) T2 : N1/N2/N3 = 14/67/17; L = 175 µH; Q_o = 145; Q_L = 50 (f = 1 MHz); V_i/V_G = −6 dB.

Fig. 7 Application circuit diagram of a AM-MW receiver with two double variable tuning capacitors; f_i = 510 to 1620 kHz (h.f.); f_i = 460 kHz (i.f.).

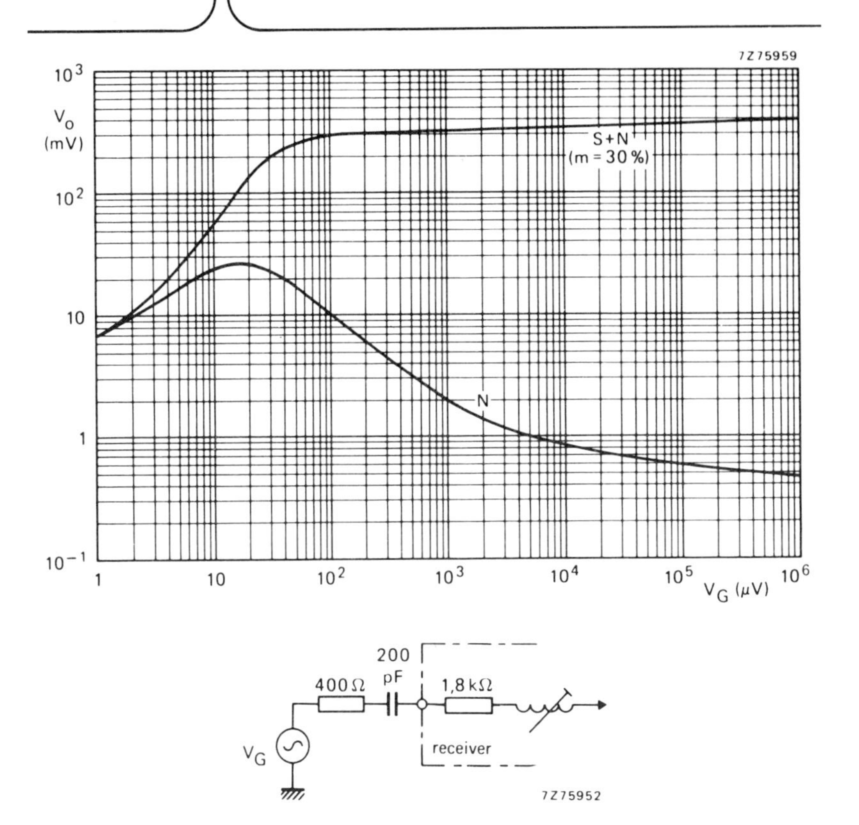

Fig. 8 A.F. output voltage as a function of the h.f. generator input voltage; f_i = 1 MHz (h.f.); f_m = 0,4 kHz.

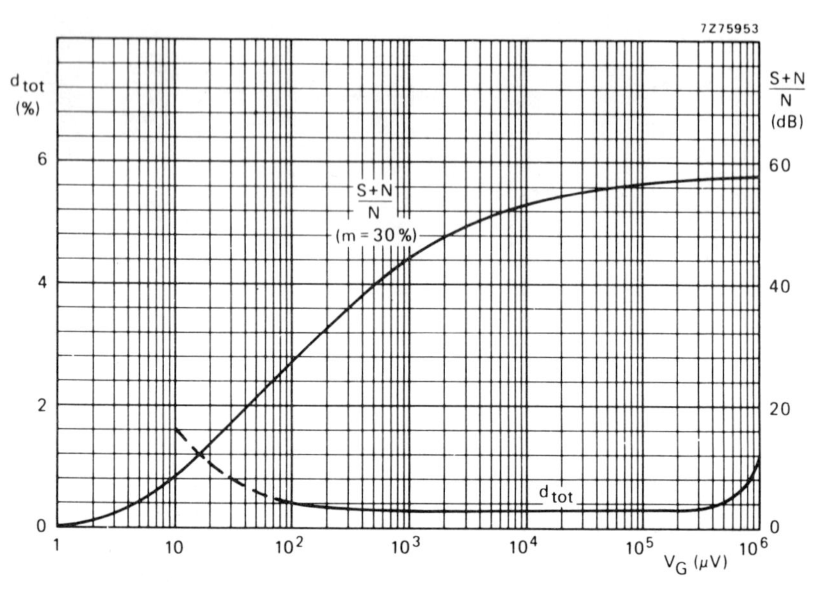

Fig. 9 Total distortion and signal plus noise-to-noise ratio as a function of h.f. generator input voltage; for d_{tot} : f_m = 0,4 kHz; m = 80%.

16-LEAD DUAL IN-LINE; PLASTIC (SOT-38)

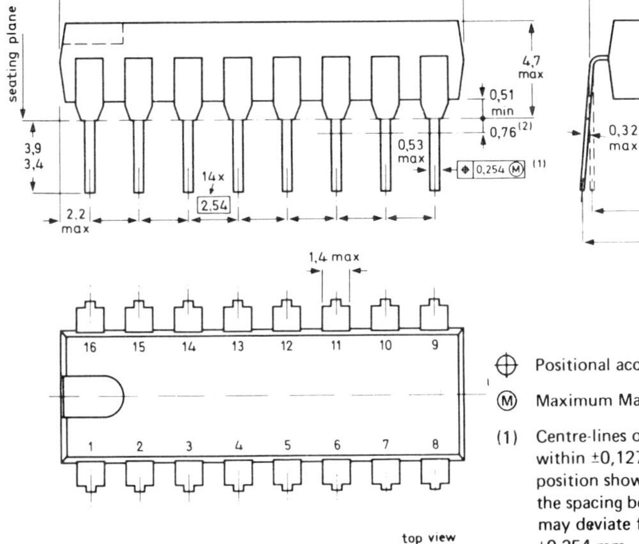

Dimensions in mm

⊕ Positional accuracy.

(M) Maximum Material Condition.

(1) Centre-lines of all leads are within ±0,127 mm of the nominal position shown; in the worst case, the spacing between any two leads may deviate from nominal by ±0,254 mm.

(2) Lead spacing tolerances apply from seating plane to the line indicated.

SOLDERING

1. By hand

Apply the soldering iron below the seating plane (or not more than 2 mm above it).
If its temperature is below 300 °C it must not be in contact for more than 10 seconds; if between 300 °C and 400 °C, for not more than 5 seconds.

2. By dip or wave

The maximum permissible temperature of the solder is 260 °C; this temperature must not be in contact with the joint for more than 5 seconds. The total contact time of successive solder waves must not exceed 5 seconds.
The device may be mounted up to the seating plane, but the temperature of the plastic body must not exceed the specified storage maximum. If the printed-circuit board has been pre-heated, forced cooling may be necessary immediately after soldering to keep the temperature within the permissible limit.

3. Repairing soldered joints

The same precautions and limits apply as in (1) above.

DESCRIPTION

The LM111 series are voltage comparators that have input currents approximately a hundred times lower than devices like the μA710. They are designed to operate over a wider range of supply voltages; from standard $\pm$15V op amp supplies down to the single 5V supply used for IC logic. Their output is compatible with RTL, DTL, and TTL as well as MOS circuits. Further, they can drive lamps or relays, switching voltages up to 50V at currents as high as 50mA.

Both the inputs and the outputs of the LM111 series can be isolated from system ground, and the output can drive loads referred to ground, the positive supply or the negative supply. Offset balancing and strobe capability are provided and outputs can be wire OR'ed. Although slower than the μA710 (200ns response time vs 40ns) the devices are also much less prone to spurious oscillations. The LM111 series has the same pin configuration as the μA710 series.

FEATURES

- **Operates from single 5V supply**
- **Maximum input bias current: 150nA (LM311 - 250nA)**
- **Maximum offset current: 20nA (LM311 - 50nA)**
- **Differential input voltage range: $\pm$30V**
- **Power consumption: 135mW at $\pm$15V**
- **High sensitivity—200V/mV**

PIN CONFIGURATIONS

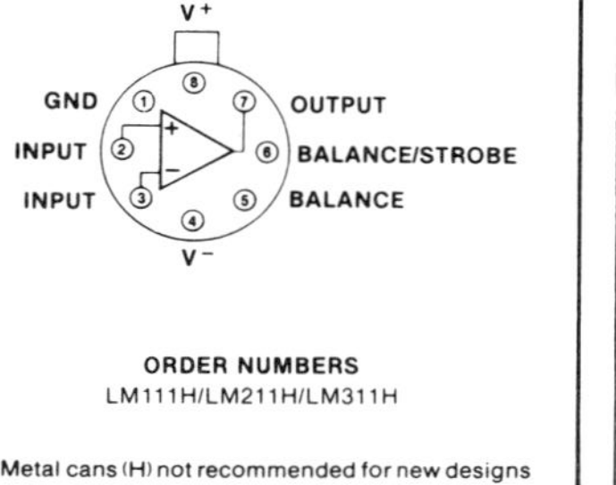

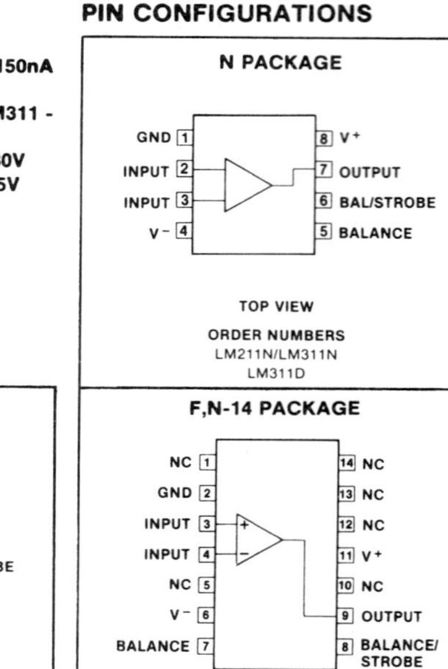

LM111 (continued)

EQUIVALENT SCHEMATIC

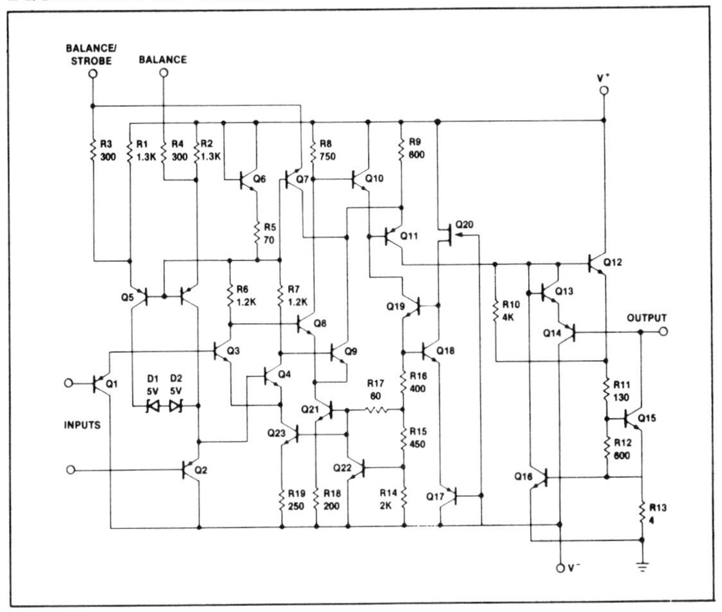

ABSOLUTE MAXIMUM RATINGS

PARAMETER	RATING	UNIT
Total supply voltage	36	V
Output to negative supply voltage:		
LM111/LM211	50	V
LM311	40	V
Ground to negative supply voltage	30	V
Differential input voltage	±30	V
Input voltage[1]	±15	V
Power dissipation[2]	500	mW
Output short circuit duration	10	sec
Operating temperature range		
LM111	-55 to +125	°C
LM211	-25 to +85	°C
LM311	0 to +70	°C
Storage temperature range	-65 to +150	°C
Lead temperature (soldering, 10sec)	300	°C

DC ELECTRICAL CHARACTERISTICS [1,2,3]

PARAMETER	TEST CONDITIONS	LM111/LM211			LM311			UNIT
		Min	Typ	Max	Min	Typ	Max	
Input offset voltage[4]	$T_A = 25°C$, $R_S \leq 50k\Omega$		0.7	3.0		2.0	7.5	mV
Input offset current[4]	$T_A = 25°C$		4.0	10		6.0	50	nA
Input bias current	$T_A = 25°C$		60	100		100	250	nA
Voltage gain	$T_A = 25°C$		200			200		V/mV
Response time[5]	$T_A = 25°C$		200			200		ns
Saturation voltage	$V_{IN} \leq -5mV$, $I_{OUT} = 50mA$ $T_A = 25°C$		0.75	1.5		0.75	1.5	V
Strobe on current	$T_A = 25°C$		3.0			3.0		mA
Output leakage current	$V_{IN} \geq 5mV$, $V_{OUT} = 35V$ $T_A = 25°C$, $I_{STROBE} = 3mA$		0.2	10		0.2	50	nA
Input offset voltage[4]	$R_S \leq 50k\Omega$			4.0			10	mV
Input offset current[4]				20			70	nA
Input bias current				150			300	nA
Input voltage range			±14			±14		V
Saturation voltage	$V+ \geq 4.5V$, $V- = 0$ $V_{IN} \leq -6mV$, $I_{SINK} \leq 8mA$		0.23	0.4		0.23	0.4	V
Output leakage current	$V_{IN} \geq 5mV$, $V_{OUT} = 35V$		0.1	0.5				μA
Positive supply current	$T_A = 25°C$		5.1	6.0		5.1	7.5	mA
Negative supply current	$T_A = 25°C$		4.1	5.0		4.1	5.0	mA

NOTES

1. This rating applies for ±15V supplies. The positive input voltage limit is 30V above the negative supply. The negative input voltage limit is equal to the negative supply voltage or 30V below the positive supply, whichever is less.
2. The maximum junction temperature of the LM311 is 110°C. For operating at elevated temperatures, devices in the TO-5 package must be derated based on a thermal resistance of 150°C/W, junction to ambient, in the N package, a thermal resistance of 162°C/W, and °C/W for the Ceramic package. The maximum junction temperature of the LM111 is 150°C, while that of the LM211 is 110°C. For operating at elevated temperatures, devices in the TO-5 package must be derated based on a thermal resistance of 150°C/W, junction to ambient. The thermal resistance of the Cerdip package is 110°C/W, junction to ambient.
3. These specifications apply for $V_S = \pm 15V$ and $0°C < T_A < 70°C$ unless otherwise specified. With the LM211, however, all temperature specifications are limited to $-25°C \leq T_A \leq 85°C$ and for the LM111 is limited to $-55°C < T_A < 125°C$. The offset voltage, offset current and bias current specifications apply for any supply voltage from a single 5V supply up to ±15V supplies.
4. The offset voltages and offset currents given are the maximum values required to drive the output within a volt of either supply with 1mA load. Thus, these parameters define an error band and take into account the worst case effects of voltage gain and input impedance.
5. The response time specified (see definitions) is for a 100mV input step with 5mV overdrive.

DESCRIPTION

CA3089 is a monolithic integrated circuit that provides all the functions of a comprehensive FM-IF system. Figure 6 is a block diagram showing the CA3089 features, which include a three-state FM-IF amplifier/limiter configuration with level detectors for each stage, a doubly-balanced quadrature FM detector and an audio amplifier that features the optional use of a muting (squelch) circuit.

The advanced circuit design of the IF system includes desirable features such as delayed AGC for the RF tuner, an AFC drive circuit, and an output signal to drive a tuning meter and/or provide stereo switching logic. In addition, internal power supply regulators maintain a nearly constant current drain over the voltage supply range of + 8 to + 18 volts.

The CA3089 is ideal for high-fidelity operation. Distortion in a CA3089 FM-IF system is primarily a function of the phase linearity characteristic of the outboard detector coil.

The CA3089 utilizes a 16-lead dual-in-line plastic package and can operate over the ambient temperature range of − 40°C to + 85°C.

FEATURES

- **Exceptional limiting sensitivity: 10μV typ. at − 3dB point**
- **Low distortion: 0.1% typ. (with double-tuned coil)**
- **Single-coil tuning capability**
- **High recovered audio: 400mV typ.**
- **Provides specific signal for control of interchannel muting (squelch)**
- **Provides specific signal for direct drive of a tuning meter**
- **Provides delayed AGC voltage for RF amplifier**
- **Provides a specific circuit for flexible AFC**
- **Internal supply/voltage regulators**

APPLICATIONS

- **High-fidelity FM receivers**
- **Automotive FM receivers**
- **Communications FM receivers**

PIN CONFIGURATION

N PACKAGE

Pin	Function	Pin	Function
1	IF INPUT	16	NC
2	IF INPUT BYPASSING	15	DELAYED AGC
3	IF INPUT BYPASSING	14	SUBSTRATE
4	FRAME	13	TUNE METER
5	MUTE CONTROL	12	MUTE LOGIC
6	AUDIO OUT	11	V+
7	AFC OUTPUT	10	REF BIAS
8	IF OUT	9	QUADRATURE INPUT

TOP VIEW
ORDER NUMBER
CA3089N

CA3089 (continued)

ABSOLUTE MAXIMUM RATINGS

PARAMETER	RATING	UNIT
DC supply voltage:		
Between terminals 11 and 4	18	V
Between terminals 11 and 14	18	V
DC Current (out of terminal 15)	2	mA
Device dissipation:		
Up to T_A = 60°C	600	mW
Above T_A = 60°C	derate linearly 6.7	mW/°C
Ambient temperature range:		
Operating	−40 to +85	°C
Storage	−65 to +150	°C
Lead temperature (during soldering):		
At distance not less than 1/32" (0.79mm) from case for 10 seconds max	+265	°C

BLOCK DIAGRAM

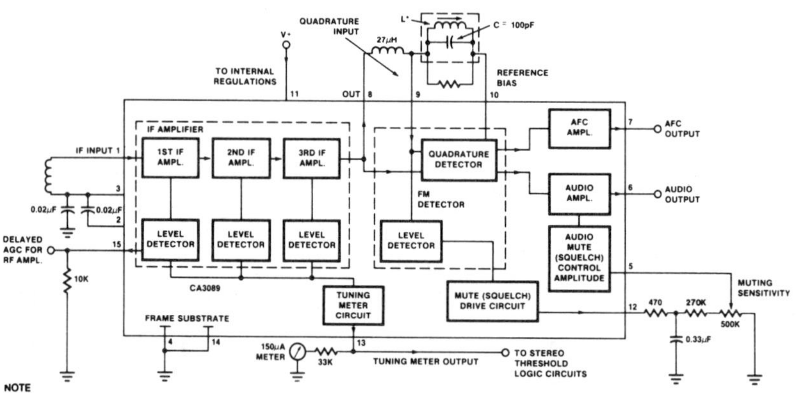

NOTE
All resistors values are typical and in ohms. $Q_0 \simeq 75$ (G.I. EX27825 or equivalent)
*L tunes with 100pF (C) at 10.7MHz

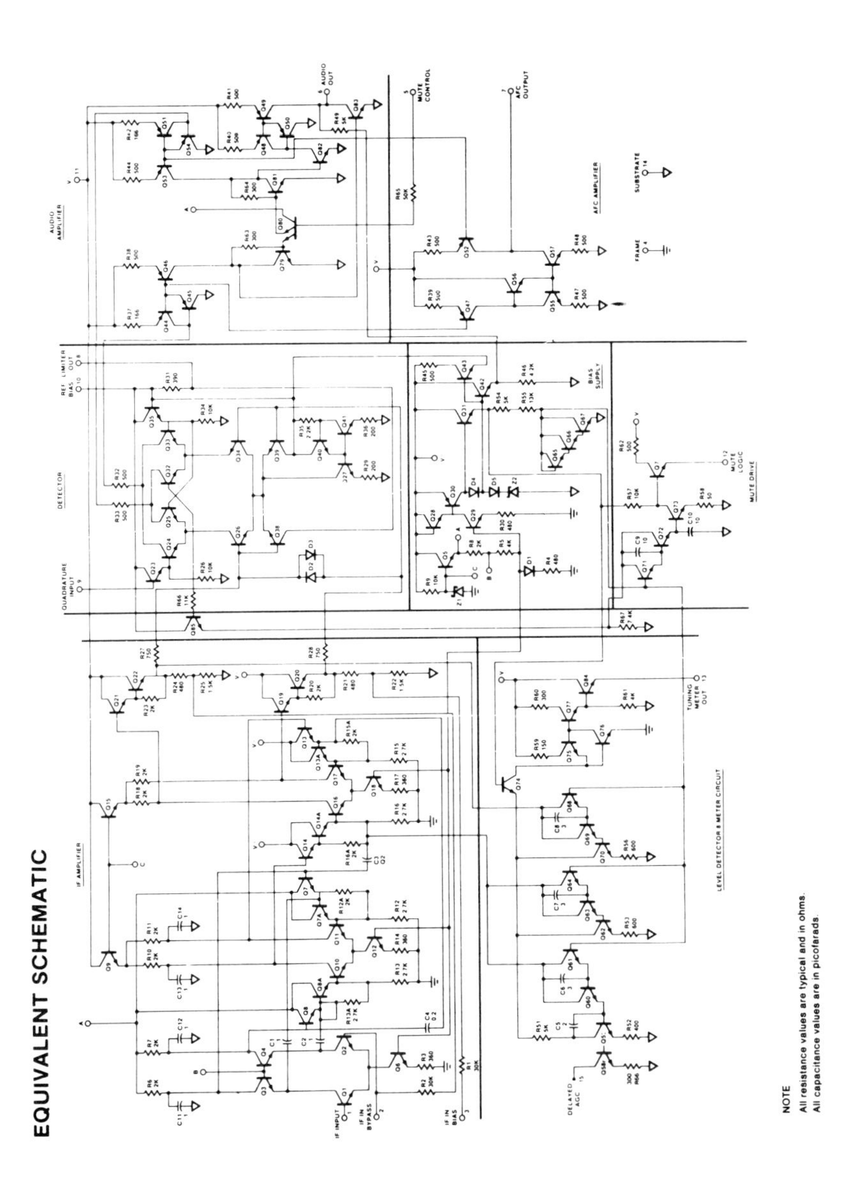
EQUIVALENT SCHEMATIC
IF AMPLIFIER
DETECTOR
AUDIO AMPLIFIER
AFC AMPLIFIER
BIAS SUPPLY
MUTE DRIVE
LEVEL DETECTOR & METER CIRCUIT
IF INPUT
IF IN BYPASS
IF IN BIAS
DELAYED AGC
QUADRATURE INPUT
REF BIAS
LIMITER OUT
AUDIO OUT
MUTE CONTROL
AFC OUTPUT
SUBSTRATE
FRAME
MUTE LOGIC
TUNING METER OUT
NOTE
All resistance values are typical and in ohms.
All capacitance values are in picofarads.

CA3089 (continued)

DC ELECTRICAL CHARACTERISTICS T_A = 25°C, V^+ = 12V unless otherwise specified.

Symbol	Parameter	Test Conditions	CA3089D2 Min	CA3089D2 Typ	CA3089D2 Max	Unit
STATIC (DC) CHARACTERISTICS						
I_{11}	Quiescent circuit current	No signal input, non-muted	16	23	30	mA
DC Voltages:[4]						
V_1	Terminal 1 (IF input)	No signal input, non-muted	1.2	1.9	2.4	V
V_2	Terminal 2 (ac return to input)	No signal input, non-muted	1.2	1.9	2.4	V
V_3	Terminal 3 (dc bias to input)	No signal input, non-muted	1.2	1.9	2.4	V
V_6	Terminal 6 (audio output)	No signal input, non-muted	5.0	5.6	6.0	V
V_7	Terminal 7 (A.F.C.)	No signal input, non-muted	5.0	5.6	6.0	V
V_{10}	Terminal 10 (dc reference)	No signal input, non-muted	5.0	5.6	6.0	V
DYNAMIC CHARACTERISTICS						
$V_{I(lim)}$	Input limiting voltage (−3dB point)[3]			10	25	μV
AMR	AM Rejection (terminal 6)[4]	V_{IN} = 0.1V, F_o = 10.7MHz, f_{mod} = 400Hz, AM Mod = 30%	45	55		dB
V_O	Recovered audio voltage (terminal 6)[3]		400	500	600	mV
Total harmonic distortion:[1]						
THD	Single tuned (terminal 6)[3]			0.5	1.0	%
THD	Double tuned (terminal 6)[4]	f_{mod} = 400Hz, V_{IN} = 0.1		0.1		%
S+N/N	Signal plus noise to noise ratio (terminal 6)[3]	Deviation = ±75kHz V_{IN} = 0.1V	60	70		dB
MU_{IN}	Mute input (terminal 5)	V_5 = 2.5V	50	70		dB
MU_{OUT}	Mute output (terminal 12)	V_{IN} = 50μV			.5	V
		V_{IN} = 0V	4.0			V
MTR	Meter output (terminal 13)	V_{IN} = 0.1V	2.5	3.5		V
		V_{IN} = 500μV	1.0	1.5		V
		V_{IN} = 0V			.7	V
AGC	Delayed AGC (terminal 15)	V_{IN} = .01V			.5	V
		V_{IN} = 10μV	4.0	5.0		V
THD	Double tuned (terminal 6)[4]	f_{mod} = 400Hz V_{IN} = 0.1		0.1		%

NOTES

1. THD characteristics and Audio Level are essentially a function of the phase and Q characteristics of the network connected between terminals 8,9, and 10.
2. Test circuit Figure 1.
3. Test circuit Figure 2.
4. Test circuit Figures 1 and 2.

TEST CIRCUITS

TEST CIRCUIT
(Using a single-tuned detector coil.)

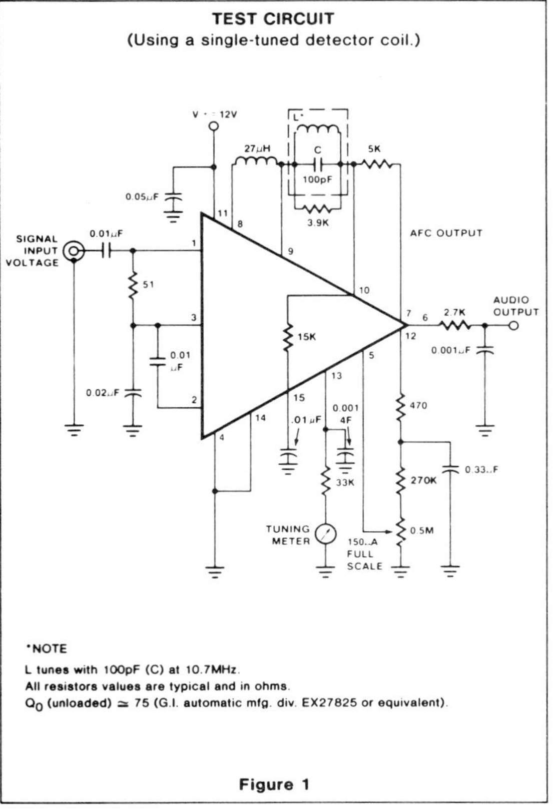

*NOTE

L tunes with 100pF (C) at 10.7MHz.

All resistors values are typical and in ohms.

Q_O (unloaded) ≃ 75 (G.I. automatic mfg. div. EX27825 or equivalent).

Figure 1

TEST CIRCUIT
(Using a double-tuned detector coil.)

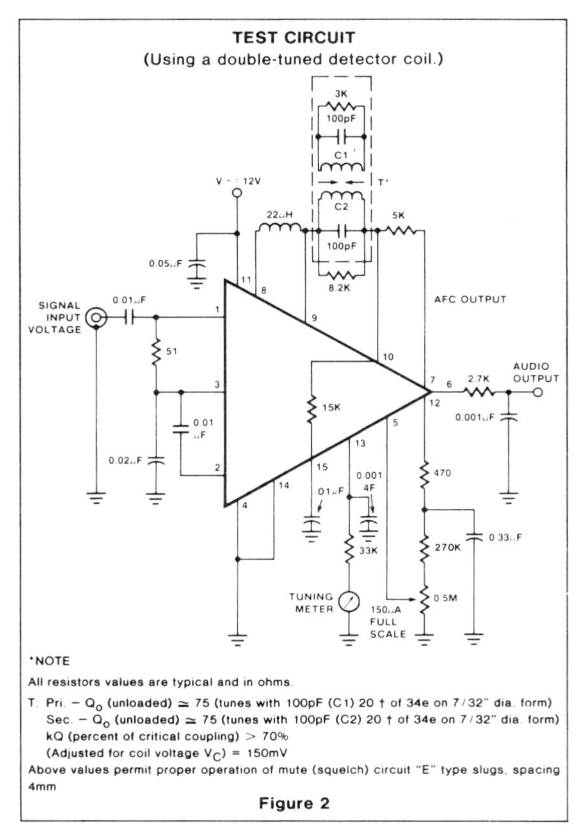

*NOTE

All resistors values are typical and in ohms.

T: Pri. – Q_O (unloaded) ≃ 75 (tunes with 100pF (C1) 20 † of 34e on 7/32" dia. form)
Sec. – Q_O (unloaded) ≃ 75 (tunes with 100pF (C2) 20 † of 34e on 7/32" dia. form)
kQ (percent of critical coupling) > 70%
(Adjusted for coil voltage V_C) = 150mV

Above values permit proper operation of mute (squelch) circuit "E" type slugs, spacing 4mm

Figure 2

ULTRA HIGH FREQUENCY OPERATIONAL AMPLIFIER — NE/SE5539

DESCRIPTION

The Signetics NE5539 is a very wide bandwidth, high slew rate, monolithic operational amplifier for use in video amplifiers, RF amplifiers, and extremely high slew rate amplifiers.

Emitter follower inputs provide a true differential high input impedance device. Proper external compensation will allow design operation over a wide range of closed loop gains, both inverting and non-inverting, to meet specific design requirements.

FEATURES

- **Gain bandwidth product: 1.2GHz**
- **Slew rate: 600V/μsec**
- **Full power response: 48MHz**
- **A_{VOL}: 50dB**

APPLICATIONS

- **Fast pulse amplifiers**
- **RF oscillators**
- **Fast sample and hold**
- **High gain video amplifiers (BW > 20MHz)**

PIN CONFIGURATION

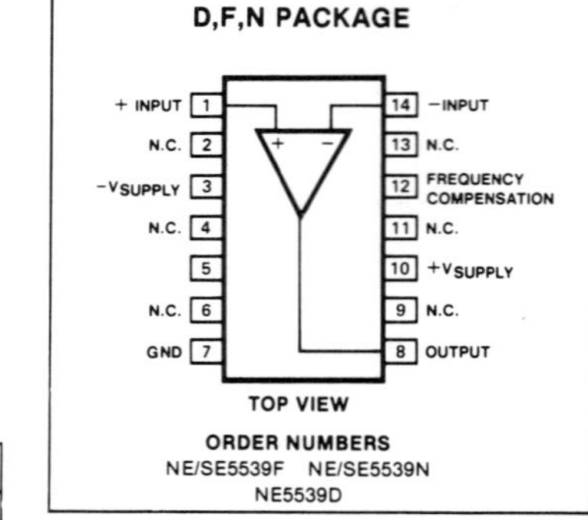

ABSOLUTE MAXIMUM RATINGS

	PARAMETER	RATING	UNIT
V_{CC}	Supply voltage	±12	V
P_D	Internal power dissipation	550	mW
T_{STG}	Storage temperature range	−65 to +150	°C
T_J	Max junction temperature	150	°C
T_A	Operating temperature range		
	NE	0 to 70	°C
	SE	−55 to +125	°C
	Lead temperature	300	°C

EQUIVALENT CIRCUIT

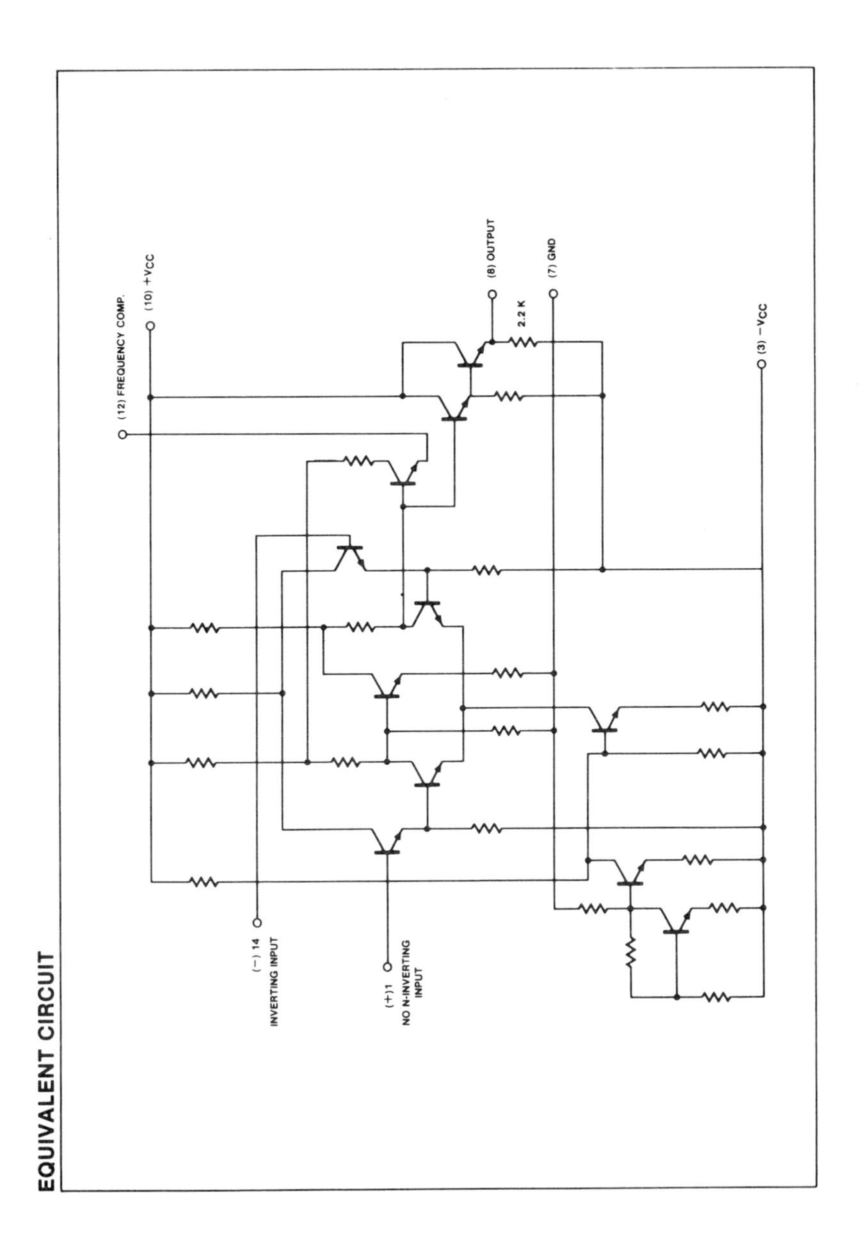

DC ELECTRICAL CHARACTERISTICS $V_{CC} = \pm 8V$, $T_A = 25°C$ unless otherwise specified

PARAMETER		TEST CONDITIONS			SE5539			NE5539			UNIT
					Min	Typ	Max	Min	Typ	Max	
V_{OS}	Input offset voltage	$V_O = 0V$, $R_S = 100\Omega$		Over temp		2	5				mV
				$T_A = 25°C$		2	3		2.5	5	
I_{OS}	Input offset current			Over temp		.1	3				μA
				$T_A = 25°C$		.1	1			2	
I_B	Input bias current			Over temp		6	25				μA
				$T_A = 25°C$		5	13		5	20	
CMRR	Common mode rejection ratio	F = 1 kHz, $R_S = 100\Omega$, $V_{CM} = 1.7V$			70	80		70	85		dB
R_{IN}	Input impedance					100			100		kΩ
R_{OUT}	Output impedance					10			10		Ω
V_{OUT}	Output voltage swing	$R_L = 150\Omega$ to GND and 470Ω to $-V_{CC}$		+Swing				+2.3	+2.7		V
				−Swing				−1.7	−2.2		
V_{OUT}	Output voltage swing	$R_L = 2k\Omega$ to GND	Over temp	+Swing	+2.3	+3.0					V
				−Swing	−1.5	−2.1					
			$T_A = 25°C$	+Swing	+2.5	+3.1					V
				−Swing	−2.0	−2.7					
$I_{CC}+$	Positive supply current	$V_O = 0$, $R_1 = \infty$		Over temp		14	18				mA
				$T_A = 25°C$		14	17		14	18	
$I_{CC}-$	Negative supply current	$V_O = 0$, $R_1 = \infty$		Over temp		11	15				mA
				$T_A = 25°C$		11	14		11	15	
PSRR	Power supply rejection ratio	$\Delta V_{CC} = \pm 1V$		Over temp		300	1000				$\mu V/V$
				$T_A = 25°C$					200	1000	

PARAMETER	TEST CONDITIONS		SE5539 Min	SE5539 Typ	SE5539 Max	NE5539 Min	NE5539 Typ	NE5539 Max	UNIT
A_{VOL} Large signal voltage gain	$V_O = +2.3V, -1.7V$ $R_L = 150\Omega$ to GND, 470Ω to $-V_{CC}$					47	52	57	dB
A_{VOL} Large signal voltage gain	$V_O = +2.3V, -1.7V$ $R_L = 2K$ to GND	Over temp							dB
		$T_A = 25°C$				47	52	57	
A_{VOL} Large signal voltage gain	$V_O = +2.5V, -2.0V$ $R_L = 2\Omega$ to GND	Over temp	46		60				dB
		$T_A = 25°C$	48	53	58				

NOTE

1. Differential input voltage should not exceed 0.25 volts to prevent excessive input bias current and common mode voltage 2.5 volts. These voltage limits may be exceeded if current limit is 10mA.

AC ELECTRICAL CHARACTERISTICS $V_{CC} = \pm 8V$, $R_L = 150\Omega$ to GND & 390Ω to $-V_{CC}$ unless otherwise specified

PARAMETER	TEST CONDITIONS	SE5539			NE5539			UNIT
		Min	Typ	Max	Min	Typ	Max	
Gain bandwidth product	$A_{CL} = 7$ $V_O = 0.1$ Vp-p		1200					MHz
Small signal bandwidth	$A_{CL} = 2$ $R_L = 150\Omega$		110					MHz
Settling time	$A_{CL} = 2$ $R_L = 150\Omega$		15					nSec
Slew rate	$A_{CL} = 2$ $R_L = 150\Omega$		600			330		V/μSec
Propagation delay	$A_{CL} = 2$ $R_L = 150\Omega$		7			10		nSec
Full power response	$A_{CL} = 2$ $R_L = 150\Omega$		48			20		MHz
Full power response	$A_V = 7$, $R_L = 150\Omega$		20					MHz
Wide band noise (RMS)	$B_W = 5MHz$, $R_S = 50\Omega$		5					μV

DC ELECTRICAL CHARACTERISTICS $V_{CC} = \pm 6V$, $T_A = 25°C$ unless otherwise specified

PARAMETERS		TEST CONDITIONS			SE5539			UNIT
					Min	Typ	Max	
V_{OS}	Input offset voltage			Over temp		2	5	mV
				$T_A = 25°C$		2	3	
I_{OS}	Input offset current			Over temp		.1	3	μA
				$T_A = 25°C$		.1	1	
I_B	Input bias current			Over temp		5	20	μA
				$T_A = 25°C$		4	10	
CMRR	Common mode rejection ratio	$V_{CM} = \pm 1.3V$, $R_S = 100\Omega$			70	85		dB
$I_{CC}+$	Positive supply current			Over temp		11	14	mA
				$T_A = 25°C$		11	13	
$I_{CC}-$	Negative supply current			Over temp		8	11	mA
				$T_A = 25°C$		8	10	
PSRR	Power supply rejection ratio	$\Delta V_{CC} = \pm 1V$		Over temp		300	1000	μV/V
				$T_A = 25°C$				
V_{OUT}	Output voltage swing	$R_L = 150\Omega$ to GND and 390Ω to $-V_{CC}$	Over temp	+ Swing	+ 1.4	+ 2.0		V
				– Swing	– 1.1	– 1.7		
			$T_A = 25°C$	+ Swing	+ 1.5	+ 2.0		
				– Swing	– 1.4	– 1.8		

AC ELECTRICAL CHARACTERISTICS $V_{CC} = \pm 8V$, $R_l = 150\Omega$ to GND and 390Ω to $-V_{CC}$ unless otherwise specified

PARAMETER	TEST CONDITIONS	SE5539			UNIT
		Min	Typ	Max	
Gain bandwidth product	$A_{CL} = 7$		700		MHz
Small signal bandwidth	$A_{CL} = 2$		120		MHz
Settling time	$A_{CL} = 2$		23		ns
Slew rate	$A_{CL} = 2$		330		V/μs
Propagation delay	$A_{CL} = 2$		4.5		ns
Full power response	$A_{CL} = 2$		20		MHz

12 Oscillators and wave-shaping circuits

12.1 Introduction

One of the most important practical requirements of electronic circuits is to produce various types of waveform. The simplest type of circuit uses simple passive components, e.g. resistors and capacitors, to 'shape' or modify any wave applied to the circuit.

In the case of oscillators – circuits designed to produce a waveform of a particular shape – a more complex circuit arrangement is required. Oscillators have many applications in electronics. In communication receivers, sine-wave oscillators are used to produce the intermediate frequency (i.f.) – this is a fixed frequency. Any input signal is 'mixed' with a low frequency to produce the i.f.; the output of the mixer is always the same frequency – this makes the design of the r.f. amplifier stage much easier. Sawtooth waves are required for such applications as oscilloscope time-bases. Square and rectangular waves (including pulses) have numerous applications in modern electronics – time-bases for microprocessor systems, etc.

Any system using waveforms of any shape will of course require test equipment which can generate suitable test waveforms.

12.2 The integrator

Figure 12.1 shows the basic arrangement for an 'integrator' circuit. Figure 12.2 shows a square-wave input to the circuit of fig. 12.1 and its output for three possible values of the circuit time constant RC.

Figure 12.2(b) shows the output when the circuit time constant RC is much less than the pulse period T. The capacitor will charge up very quickly and remain at the same level as the input until the input falls to zero. The capacitor will then discharge very quickly, and the output is almost identical to the input.

Figure 12.2(c) shows what the output will look like when the circuit time constant is approximately the same as the pulse period. In a time equal to the time constant, a capacitor will charge to 63.2% of the applied voltage; thus, by the time the input wave has fallen to zero after the pulse, the capacitor will have charged to approximately $\frac{2}{3}V$. It will then discharge, again with the same time constant.

In fig. 12.2(d), when the circuit time constant is much greater than the period of the input wave, the capacitor will hardly have started to charge before the input has fallen back to zero, producing a waveform as shown.

The condition in fig. 12.2(d) produces an output which is an approximation to the mathematical integral of the input waveform.

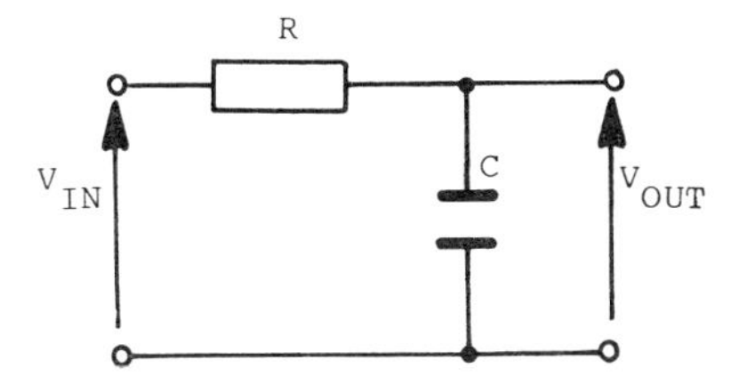

Fig. 12.1 Basic integrator

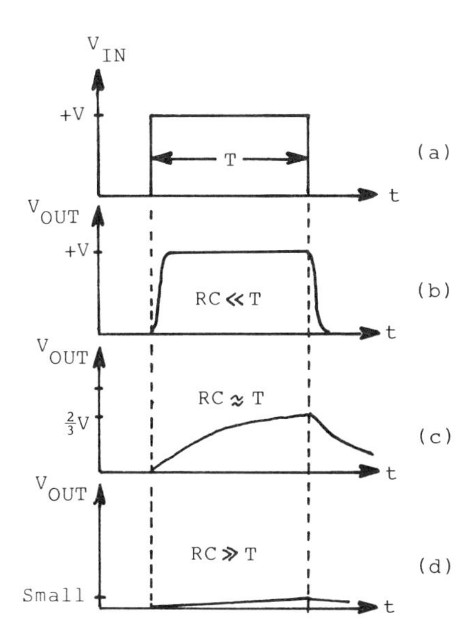

Fig. 12.2 Waveforms for fig. 12.1

12.3 The differentiator

Figure 12.3 shows the basic arrangement for the differentiator circuit. Figure 12.4 shows a square-wave input to the circuit of fig. 12.3 and the output waveforms for various circuit time constants.

Figure 12.4(b) shows the output voltage when the circuit time constant is much less than the period of the input wave. As soon as the potential on the left-hand plate of capacitor *C* rises to *V* then so must that on the right-hand plate, because the capacitor cannot change its charge instantaneously. However, the capacitor charges up quickly and the output falls exponentially to ground. When the input returns to zero, the left-hand plate of *C* is grounded. Immediately before this happening, the right-hand plate of the capacitor was at *V* volts less than the left-hand. As the

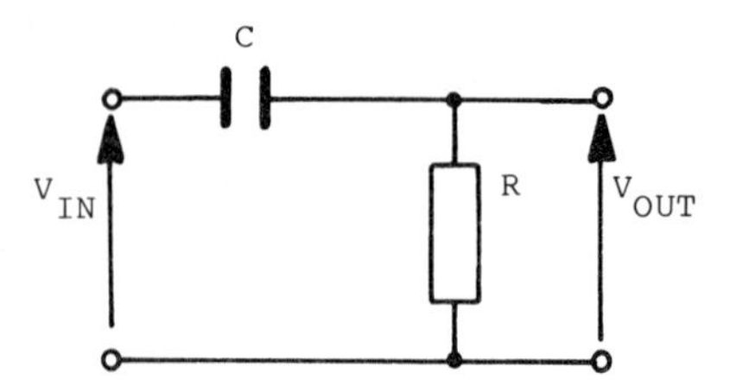

Fig. 12.3 Basic differentiator

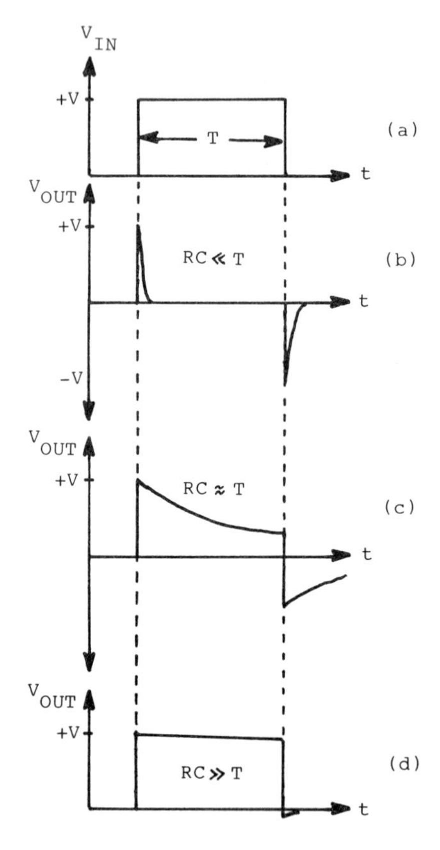

Fig. 12.4 Waveforms for fig. 12.3

capacitor cannot change its charge instantaneously, the output falls to V volts below ground until the capacitor discharges rapidly and the output 'rises' to zero.

In fig. 12.4(c), when the circuit time constant is approximately equal to the pulse period, the output will rise initially to V and in time T the capacitor will charge up to $\frac{2}{3}V$; hence the output will fall to about $\frac{1}{3}V$.

In fig. 12.4(d), when the circuit time constant is much larger than T, the output will be almost the same as the input.

The condition in fig. 12.4(b) produces an output which is an approximation to the mathematical differential of the input waveform.

The actual values of R, C, and T will determine how closely the outputs approach the ideal waveforms. Because of problems with stability, differentiators are not very common.

12.4 Feedback oscillators

As mentioned in chapter 7 on feedback, positive feedback has its main use in oscillators. Feedback is generally achieved via an inductor–capacitor network or a resistor–capacitor network. All oscillators – as opposed to amplifiers which oscillate due to unwanted feedback – have three requirements for controlled oscillations to occur:

i) positive feedback, i.e. the component fed back must be in phase with the input signal;
ii) the overall gain of the system must be greater than unity, or oscillations will die away;
iii) a frequency-determining network must be used to ensure that the frequency of oscillation is the desired one.

Such a system is shown in block-diagram form in fig. 12.5.

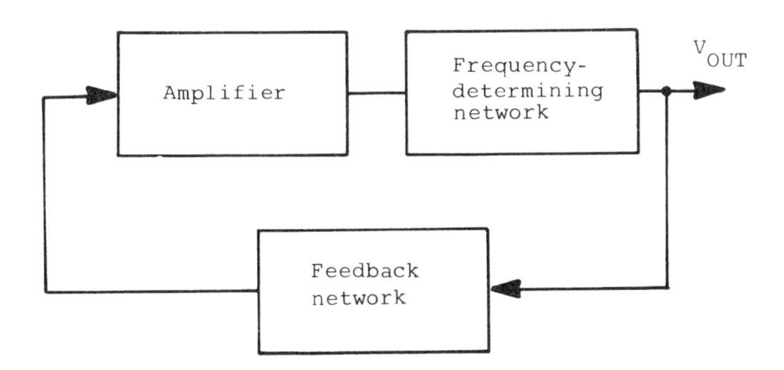

Fig. 12.5 Principle of an oscillator

12.5 Tuned-circuit oscillators

Basic tuned-collector and tuned-drain circuits are shown in fig. 12.6. In these, a parallel-tuned L–C circuit provides the frequency-determining circuit. If the resistance of the L–C circuit is very low, the frequency of oscillation, f_r, approximates to

$$f_r = \frac{1}{2\pi\sqrt{(LC)}} \qquad 12.1$$

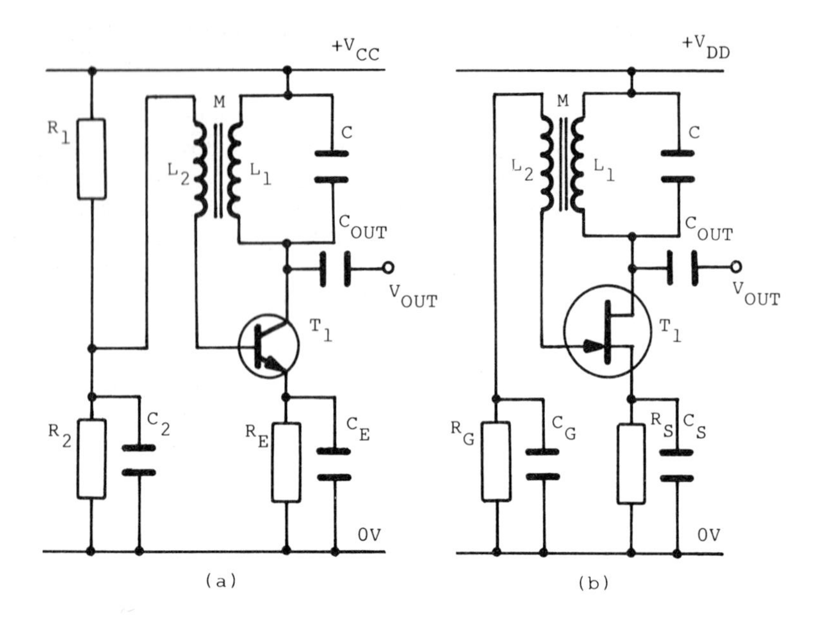

Fig. 12.6 Tuned-collector and tuned-drain oscillators

This is the resonant frequency of the parallel-tuned circuit, which presents a very high dynamic *resistive* load to the amplifier. Since the load is resistive, the voltage across the tuned circuit is exactly 180° out of phase with the base input. If the secondary of the transformer is connected to the base in such a way as to give a further 180° phase shift, the total phase shift of the feedback network is 360°, so at this frequency oscillations will occur.

12.5.1 Operation

Both circuits operate in a similar manner. On switch-on, due to the low d.c. resistance of the coil a large collector (or drain) current will flow, causing an induced e.m.f. in the coil which charges up the capacitor C. As C charges up, collector (or drain) current falls and ceases when C is fully charged. The oscillatory action of the L–C circuit will then continue, drawing a small current from the supply to replace any small resistive losses in the circuit.

Transistor circuit, fig. 12.6(a) Initial base-current bias is provided by the potential-divider network R_1 and R_2. Resistor R_E provides thermal stabilisation and is bypassed for a.c. by C_E. Positive feedback is obtained from the mutual coupling between L_1 and L_2, the feedback signal being developed across L_2; therefore, to prevent the initial d.c. bias being affected, R_2 is decoupled for a.c. – i.e. one end of L_2 is grounded to a.c.

Oscillations build up and are self-sustaining. The oscillator can be biased into class A, giving the purest sine-wave output, or class C for efficiency and rapid build-up of oscillation.

FET circuit, fig. 12.6(b) Bias is achieved by the R_G–C_G parallel network. Resistor R_G is chosen to be a high value, to ensure that the $R_G C_G$ time constant is long compared to the period of the required output signal. This allows C_G to charge up to the peak value of the gate voltage, which then acts as the gate bias driving the gate positive for a short time at the peak of the signal swing, allowing a small drain current to replace losses in the tuned circuit. The circuit operates in class C.

On switch-on there is no bias until C_G charges up, so initially a large drain current flows, charging up C_G, adjusting the bias automatically, reducing the FET gain, and allowing the amplitude to stabilise itself at a value which ensures that the loop gain is just sufficient to maintain oscillations.

The circuit is stable and highly efficient (class C), but, if oscillations do not start, then a high drain current flows which could damage the device. The circuit requires a relatively long time for bias to build up.

12.6 Hartley oscillator

Another type of L–C oscillator is the 'Hartley' oscillator, which occurs in a number of variations. Figure 12.7(a) shows one form. The circuit consists of the basic common-emitter amplifier with R_1, R_2, and R_E

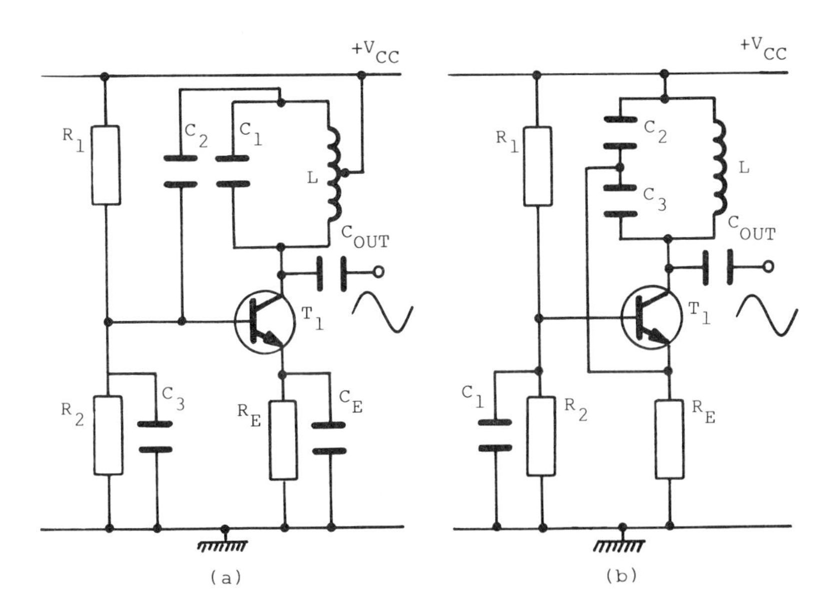

Fig. 12.7 (a) One form of Hartley oscillator, (b) Colpitt's oscillator

providing initial bias and stabilisation. The collector load is the parallel-tuned circuit L–C_1. The collector and base are connected to opposite ends of the tuned circuit.

Positive feedback is obtained from the centre tap on the inductor and is fed back directly to the emitter, which is always in phase with the collector. Remember, the power supply and C_E are effective short circuits to a.c. The amplitude is self-limiting for, if the voltage at the emitter rises too high, V_{BE} will fall. This tends to distort the output, but the L–C circuit smooths the output. The output is sinusoidal at a frequency f_r given by

$$f_r = \frac{1}{2\pi\sqrt{(LC)}} \qquad 12.2$$

If the capacitor C_1 is replaced by two capacitors and the feedback to the emitter is obtained from the centre tap on the capacitors, a *Colpitt's oscillator* is formed. Figure 12.7(b) shows a typical circuit.

12.7 *R–C* oscillators

At low oscillation frequencies – below about 50 kHz – the physical size and cost of the inductor in the L–C oscillators start to become a problem, and the inductor is prone to picking up interference. As a result, oscillators using R–C combinations are preferred.

12.7.1 Phase-shift oscillator

The basic circuit of a phase-shift oscillator is shown in fig. 12.8. The amplifier is biased and stabilised in the normal way for d.c. Since the collector output is 180° out of phase with the base, a 'phase-shifting' network consisting of three R–C combinations is provided. If the values of

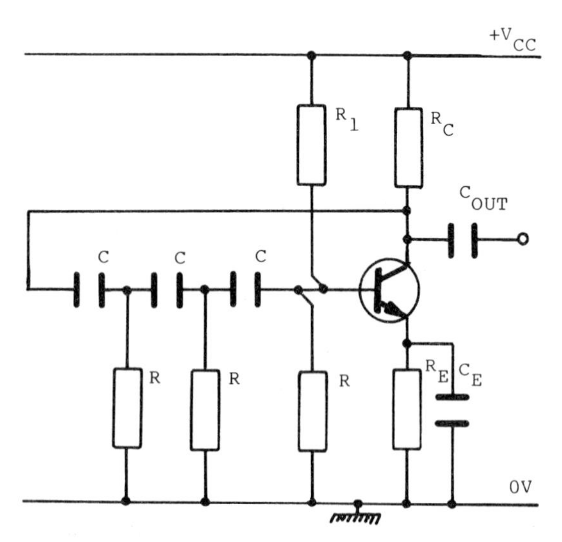

Fig. 12.8 Phase-shift oscillator

R and C are suitably chosen, the output voltage of each of the three networks will be shifted 60°, giving a total phase shift of 180° and the signal is now in phase with the base input. A minimum of three R–C networks is required – in theory, ϕ in fig. 12.9 could be 90°, requiring only two networks, but this would require R to be almost zero which would effectively short the fed-back signal to ground, and in any case the final resistor in the network has a dual function: it is also part of the d.c. biasing network.

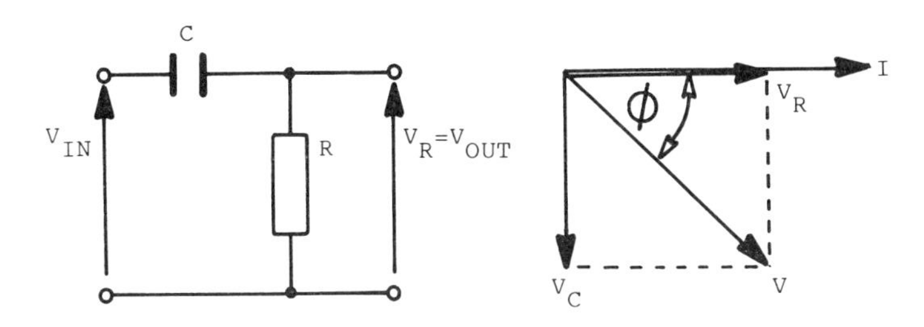

Fig. 12.9 Phase shift across an R–C network

Frequency of oscillation can be shown to be

$$f_r = \frac{1}{2\pi RC\sqrt{6}} \qquad 12.3$$

It is difficult to obtain variable-frequency outputs using this method, due to the practical requirement that all capacitors or all resistors be varied simultaneously – the resistors cannot be used because one is a d.c. bias component, and variable capacitors above 500 pF are difficult to obtain.

12.8 Wien-bridge oscillator

This circuit provides easy control of the frequency of oscillation. Figure 12.10 shows the principle of operation. An op-amp or a common-emitter amplifier with an *even* number of stages may be used in conjunction with the Wien-bridge network, which consists of a series and parallel R–C combination.

If the resistors in the feedback network and the capacitors are equal, the voltages V_{OUT} and V_{FB} can be shown to be in phase at only one frequency, given by

$$f_r = \frac{1}{2\pi RC} \qquad 12.4$$

Provided the overall gain of the amplifier is greater than 3, oscillations will be maintained.

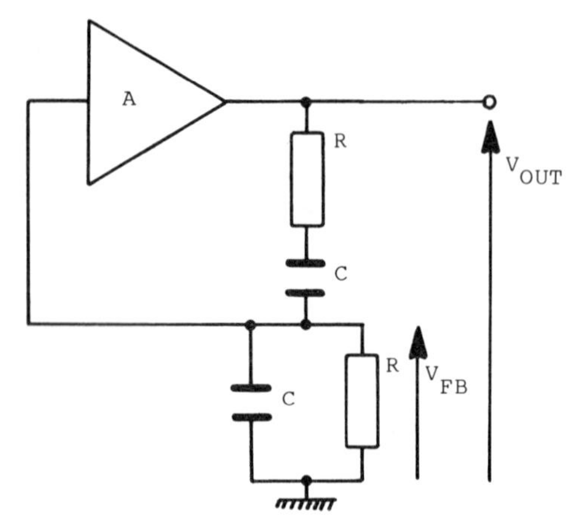

Fig. 12.10 Wien-bridge oscillator

12.9 Frequency stability

The frequency stability of an oscillator is its ability to maintain a constant frequency over a prolonged period.

As with all semiconductor-based devices, temperature will be the main factor affecting short-term stability, with ageing having a longer-term effect. Variations in supply voltage have only a minimal effect. Any changes in the load on the oscillator can have a marked effect; this can be reduced by using a 'buffer amplifier' – a unity-gain high-input-impedance amplifier – as shown in fig. 12.11.

If high stability is required – as in communication receivers, transmitters, and microprocessor systems – *crystal oscillators* must be used. Crystals are materials such as quartz which resonate mechanically when a voltage is applied across the faces of the crystal – the 'piezo-electric' effect. The electric force produces a mechanical deformation which can then be used to set up an electric field repeating the process. The resonant frequency is similar to that of a parallel-tuned circuit, having a high Q factor (the current magnification of the circuit).

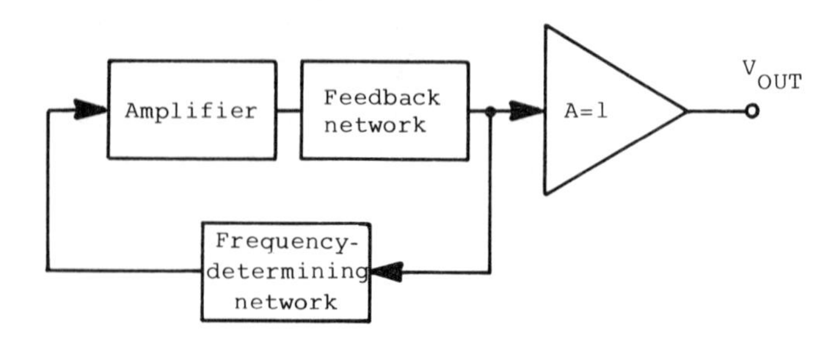

Fig. 12.11 Use of a buffer amplifier

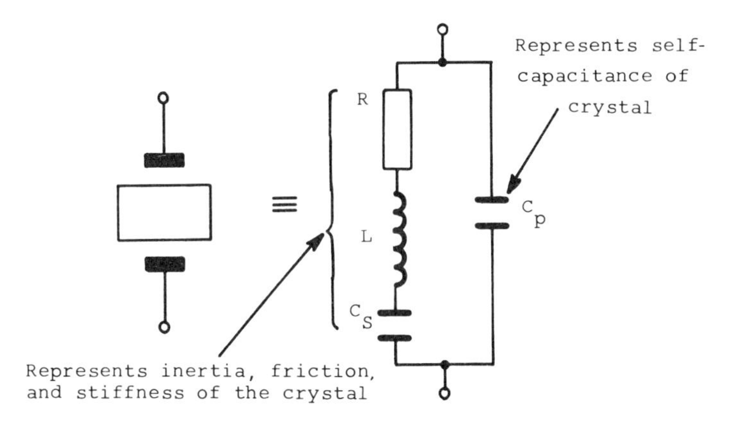

Fig. 12.12 Crystal symbol and electrical equivalent circuit

The symbol and electrical equivalent circuit are shown in fig. 12.12. The frequency is predetermined by the physical characteristics of the device, which can be controlled easily. Typical frequencies range from 4 kHz to 48 MHz with temperature stability very good, typically in the range 10 parts in 10^6 to 100 parts in 10^6 for commercially available devices. Oscillators with frequency stabilities of 1 part in 10^8 can be obtained using temperature control.

Figure 12.13 shows the tuned-collector circuit of fig. 12.6(a) modified to achieve stability using a crystal.

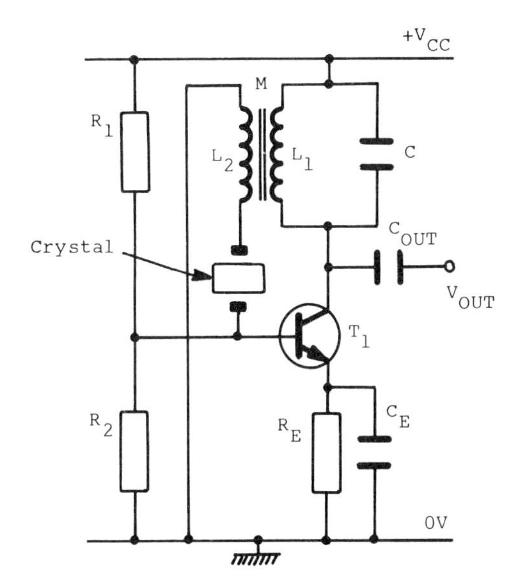

Fig. 12.13 Crystal-controlled oscillator

12.10 Multivibrators

Multivibrators are a special class of oscillators belonging to the family called 'relaxation oscillators' – these oscillators produce non-sinusoidal outputs. Multivibrators have two possible output states – the output remains constant in one state determined by an R–C timing network, then switches to the other state for a period determined by another R–C network. They produce square, rectangular, and pulse outputs.

There are three basic types of multivibrator:

i) *Astable or free-running* This provides a rectangular or square-wave output at a fixed or variable pulse-repetition frequency and mark-to-space ratio (see section 12.12). Mark-to-space ratio is the ratio of the time periods t_2 and t_1 in fig. 12.15, i.e. t_2:t_1.
ii) *Monostable or 'one-shot'* This remains in one state until it is triggered into the other state, where it remains for a fixed time period determined by an R–C network.
iii) *Bistable or flip–flop* This remains in either of its two stable states until a trigger signal is applied, when it changes to the other state where it then remains until another trigger signal is applied.

The circuits and waveforms of discrete multivibrators are described next. Practically, however, there is little call for using discrete construction since stable circuits carrying out the three functions are easily constructed using standard commercially available devices in the TTL or CMOS ranges. (TTL is the transistor–transistor logic family, available in the '7400' series of devices, and is made using bipolar-transistor technology. CMOS is a family of devices – the '4000' series – which carry out similar functions to the TTL devices but are based on MOSFET technology.)

12.10.1 Astable multivibrator

The basic astable-multivibrator circuit is shown in fig. 12.14 and the output waveforms in fig. 12.15. The transistors are in common-emitter mode, and the output can be taken from either collector.

One transistor, say T_1, will always switch on faster than the other. As T_1 switches on, its collector potential V_{C1} falls. This fall is fed to the base of T_2, causing it to switch off. This causes the collector potential V_{C2} to rise, which is fed back to the base of T_1, reinforcing the process. While T_2 is off – provided the time constant R_4C_2 is shorter than R_2C_1 – C_2 will charge up to the supply voltage. Eventually, as C_1 charges up through R_2, the base potential of T_2 will rise sufficiently to allow T_2 to switch on (base about 0.6 V). T_2 saturates quickly and its collector voltage falls to zero. This means that the right-hand plate of C_2 is now at ground potential and, since the voltage across C_2 is equal to the supply voltage V_{CC}, the left-hand plate and hence V_{B1} must be driven negative by V_{CC} which switches T_2 off. C_1 now charges up to quickly to V_{CC} through R_1. Capacitor C_2 is being

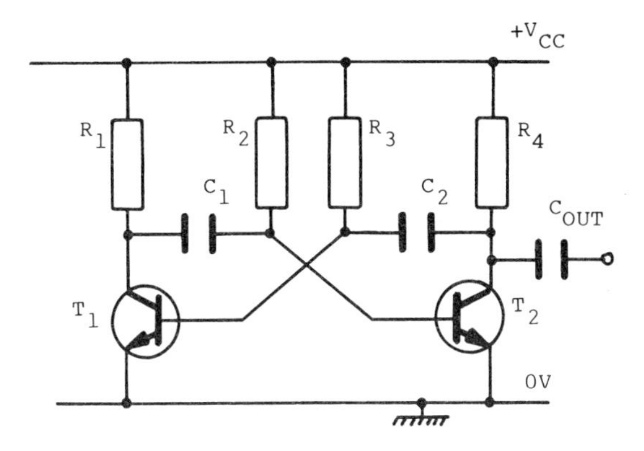

Fig. 12.14 Astable multivibrator

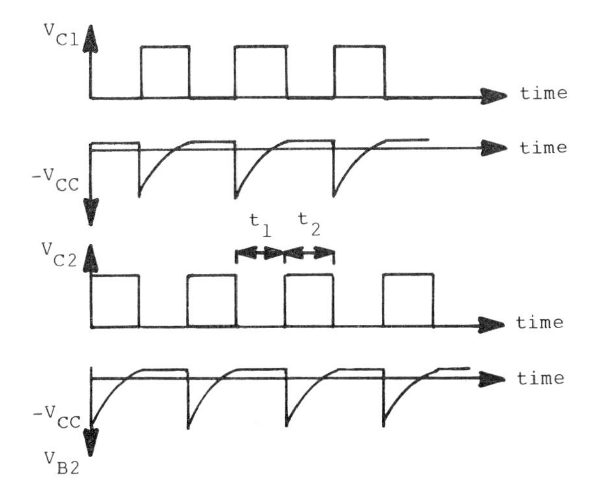

Fig. 12.15 Waveforms for circuit of fig. 12.14

charged from $-V_{CC}$ to $+V_{CC}$ through R_3,* allowing the base potential of T_1 to rise. When it has risen to about 0.6 V above ground, T_1 switches on and the process repeats itself. The relative times can be shown to be approximately

* Some explanations say the capacitor 'discharges', since it was charged up to $+V_{CC}$. If this were so, then it would take about 5 times the time constant for the voltage across the capacitor to fall to zero before T_1 switched on again. This is *not* the case. The actual time can only be calculated using an assumption that the capacitor is charging from $-V_{CC}$ to $+V_{CC}$, i.e. it is being *recharged* in the opposite direction.

$$t_1 = 0.7\, C_1 R_2 \qquad 12.5$$

$$t_2 = 0.7\, C_2 R_3 \qquad 12.6$$

The circuit does not require an external signal input. It can be used to generate square waves for clocks, timing sequences, SCR firing, etc. The mark-to-space ratio can be varied by varying R_2, R_3, C_1, or C_2.

If the time constants C_1R_1 and C_2R_4 are not small compared with C_1R_2 and C_2R_3, the rising edge of the collector waveforms will be distorted as shown in fig. 12.16.

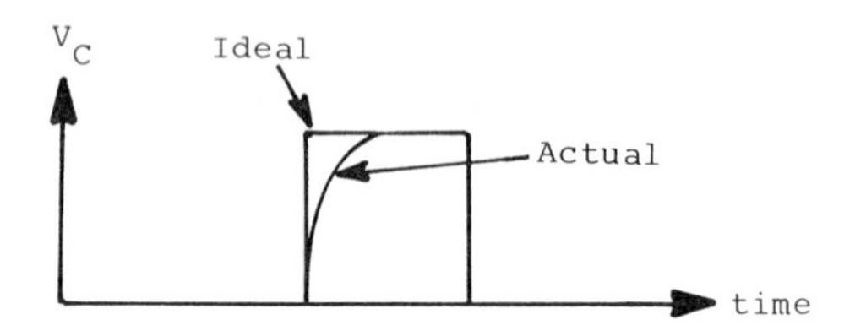

Fig. 12.16 Distortion of leading edge of collector waveform

12.10.2 Monostable multivibrator

The basic monostable-multivibrator circuit is shown in fig. 12.17 and the output waveforms in fig. 12.18.

The circuit is stable in one state. T_1 is biased to cut-off by $-V_{BB}$, so T_2 is on. If a positive pulse at least 0.6 V greater than $-V_{BB}$ is applied to the

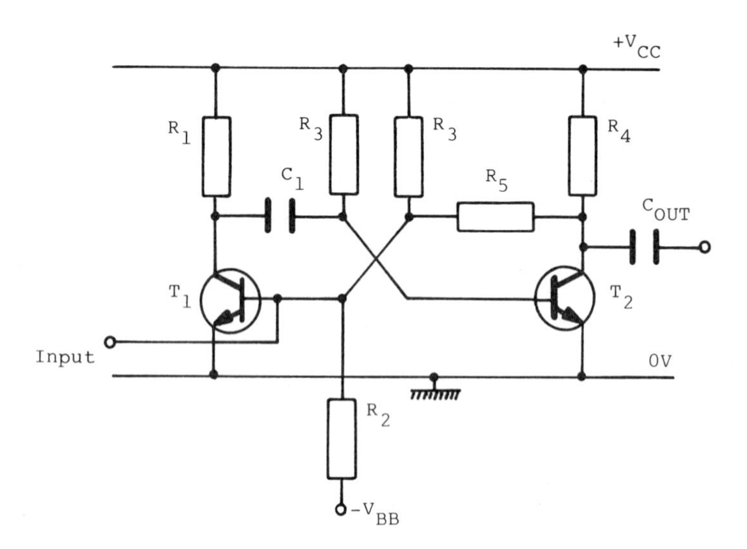

Fig. 12.17 Monostable multivibrator

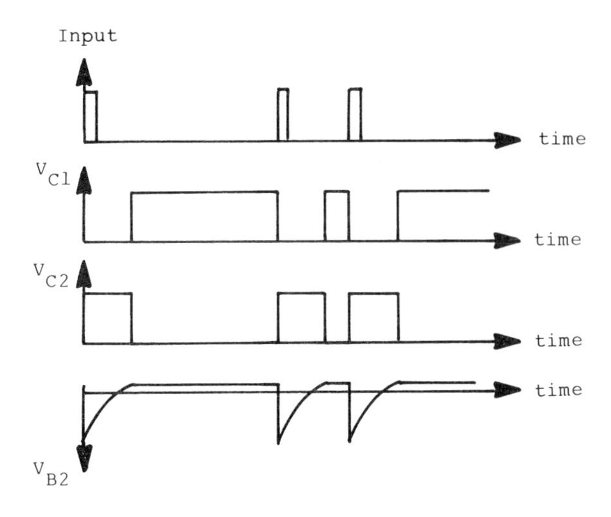

Fig. 12.18 Waveforms for the circuit of fig. 12.17

input, T_1 will switch on and its collector potential falls to zero, causing the base of T_2 to be driven to $-V_{CC}$ by the volt drop across C_1. T_2 will remain off for a time $t = 0.7\ C_1R_3$ before switching back on. Provided the input pulse has been removed, the potential-divider network R_2, R_4 and R_5 will ensure that T_1 switches off.

The circuit gives one output pulse for each input pulse. It can be used to convert pulses of one shape to square pulses of the same frequency, to reconstitute distorted pulses, to delay pulses, or to convert the voltage levels of signals.

12.10.3 Bistable multivibrator

The basic bistable-multivibrator circuit is shown in fig. 12.19. The circuit will remain in whichever state it is in until an input pulse is applied – the pulse will be guided to the correct base by the potential-divider networks R_1, R_3, R_6 or R_2, R_4, R_5. The circuit will then switch to the other state and remain there until another pulse is applied. It is used as the basic static-memory device in calculators and computers, as a divider or counter.

12.11 Multivibrator synchronisation

Some applications which use multivibrators require that the cycle starts in synchronisation with an external signal, e.g. the line and field synchronisation pulses in TV signals are used to 'lock' the picture. This is achieved by detecting the pulses and superimposing them on the base waveform of one of the transistors in the astable multivibrator of fig. 12.14 as shown in fig. 12.20. If the combined value of the 'synch-pulse' and normal base potential is greater than $+0.6$ V, the transistor will be switched on or 'synchronised' with the signal.

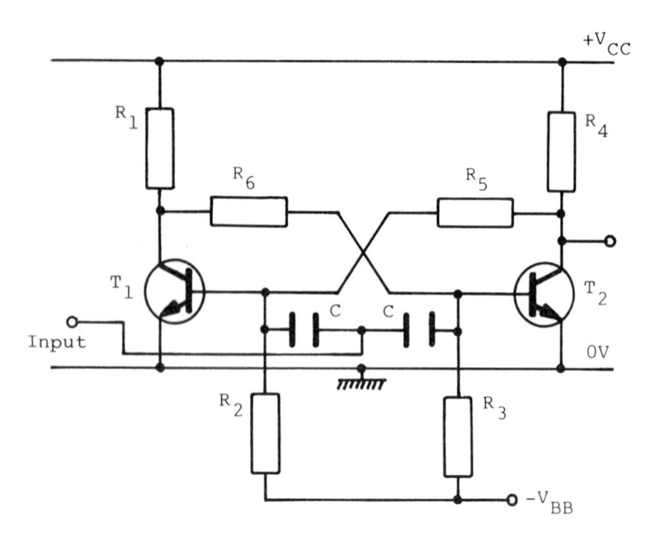

Fig. 12.19 Bistable multivibrator

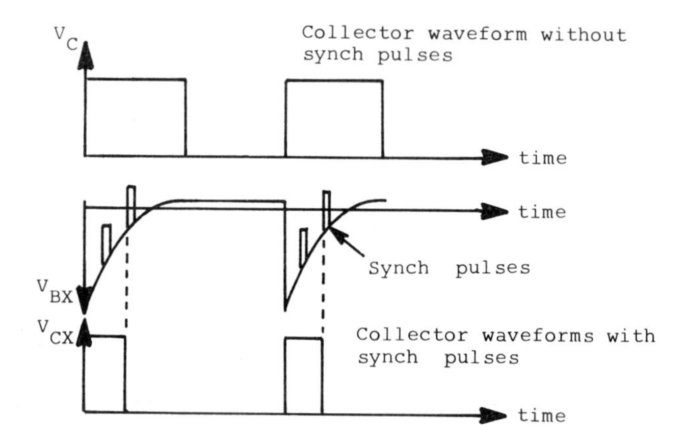

Fig. 12.20 Effect of synch pulses on collector waveform

12.12 Waveforms

Circuits which produce square, rectangular, or pulse waveforms do not in practice produce perfect waves – an actual waveform differs from the ideal because of the effects of inductance and capacitance which all circuits posses.

A practical waveform is shown in fig. 12.21(a). The amount of distortion is usually indicated by measuring the 'rise time' t_R and the 'fall time' t_F.

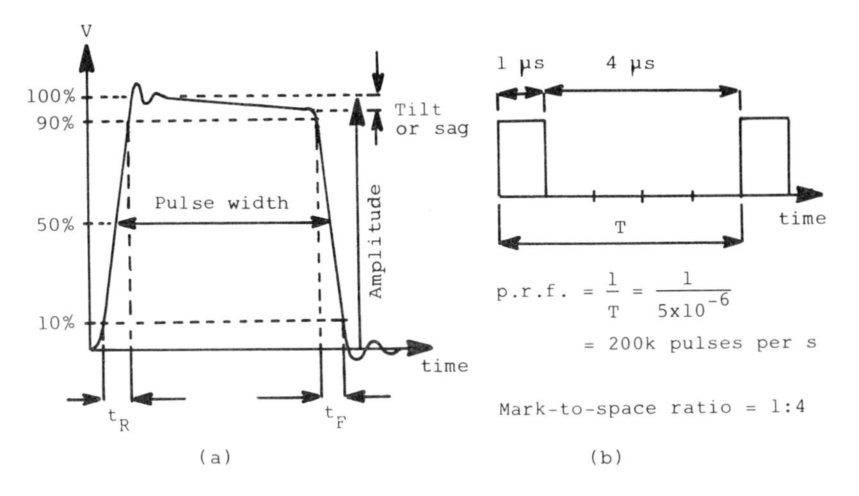

Fig. 12.21 (a) Waveform distortion, (b) example of mark-to-space ratio

The rise time is the time for the voltage to rise from 10% to 90% of its final value.

The fall time (decay time) is the time for the voltage to fall from 90% to 10% of its amplitude.

The pulse width is generally measured between the 50% points on the waveform (although it may be quoted between the 10% or 90% points).

Many waveforms also have small amounts of 'overshoot' and 'undershoot', accompanied by a small amount of 'ringing' as the waveform settles down. A practical waveform will also decay slightly, known as 'tilt' or 'sag'. Many of these differences from the ideal are hardly noticeable in operation.

The term 'frequency' does not completely define rectangular or pulse waveforms – two more quantities need to be specified:

i) *pulse-repetition frequency* (p.r.f.), i.e. the number of pulses per second;
ii) *mark-to-space ratio*, i.e. the ratio between the pulse width (or mark) and the time between the pulses. An example is shown in fig. 12.21(b).

12.13 Some practical circuits

12.13.1 Schmitt-trigger oscillator

One of the simplest oscillators which can be made for use with digital circuits is the Schmitt trigger. Figure 12.22(a) shows the basic circuit. A Schmitt-trigger device is one which requires the input voltage to reach a particular 'threshold' value before any change in the output takes place – the change in the output state is then very quick. This produces very good

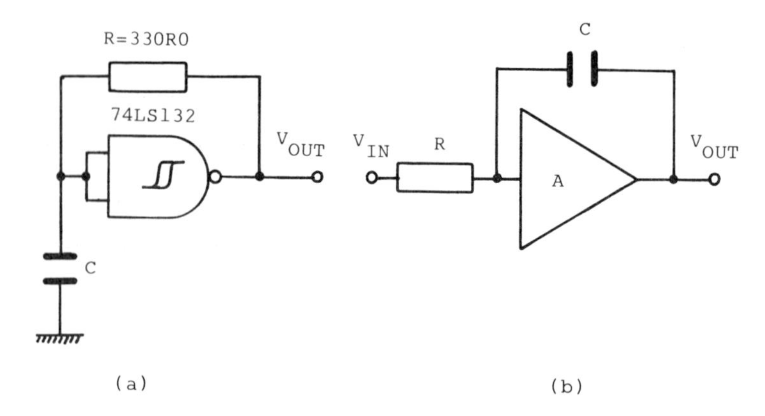

Fig. 12.22 (a) Simple Schmitt-trigger oscillator, (b) Miller integrator

'square' edges to waveforms. The 'hysteresis loop' in the NAND-gate symbol in fig. 12.22(a) indicates that the device is a Schmitt input gate. A typical device is the 74LS132 gate of the TTL logic family – this is a quad (four) two-input NAND gate.

The external resistor, R, must be kept small – typically 390 Ω. Frequencies from very low (< 1 Hz) up to 10 MHz are possible, the frequency depending on the value of R and C. For $R = 390\ \Omega$,

$$f \approx \frac{1}{390\ C}$$

12.13.2 Blocking oscillator

At a first glance, the blocking-oscillator circuit, fig. 12.23(a), looks like a sinusoidal oscillator, but it is in fact a relaxation oscillator. The coupling between the primary and secondary windings of the transformer is so tight that the transistor is biased off, or 'blocked', until the bias voltage on capacitor C discharges. Determination of the pulse-repetition frequency for this circuit is extremely complex and is dependent on the transformer and the values of R and C. Typical waveforms are shown in fig. 12.23(b). D_1 is a 'flywheel' diode – this is a diode placed across inductive loads such as motors and relay coils so that it is normally reverse-biased and so does not conduct; however, if an e.m.f. is induced by the current changing in the inductance, the diode will provide a path for any current generated.

The blocking-oscillator circuit can be used as a pulse generator in digital circuits.

12.13.3 Miller integrator

The simple integrator circuit of fig. 12.1 can be modified by the addition of an op-amp as shown in fig. 12.22(b). The effect of the op-amp is to

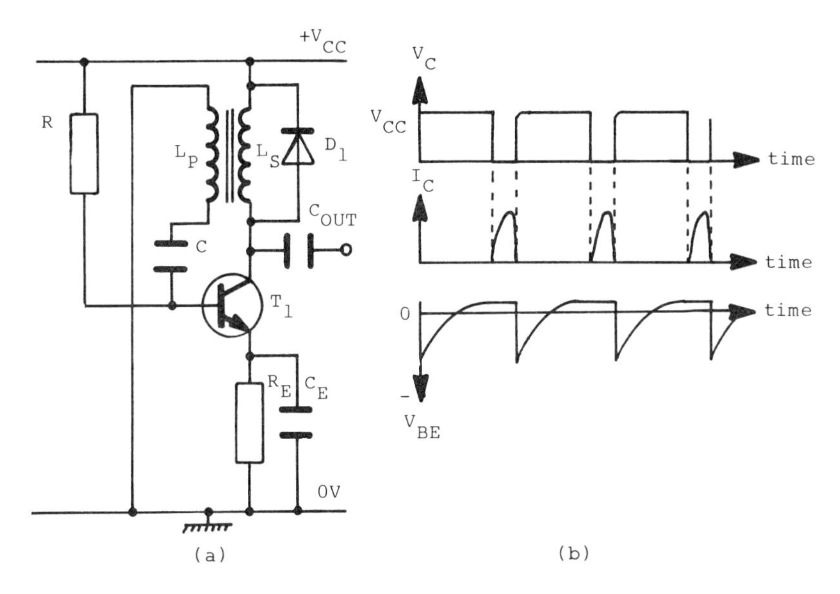

Fig. 12.23 Blocking oscillator and waveforms

increase the effective circuit capacitance to $C \times A$, which gives a more linear output waveform.

Problems

12.1 For the pulse waveform in fig. 12.24(a), calculate (a) the pulse-repetition frequency (p.r.f.), (b) the mark-to-space ratio.

12.2 For the pulse in fig. 12.24(b), calculate (a) the rise time, (b) the fall time, (c) the pulse width.

12.3 The phase-shift oscillator of fig. 12.25 has variable capacitors with a range 100 pF to 500 pF. The resistors R are 20 kΩ. Calculate the frequency range of the oscillator.

12.4 If the capacitor C in the tuned-collector oscillator of fig. 12.6(a) is variable from 100 pF to 500 pF and the maximum frequency of oscillation

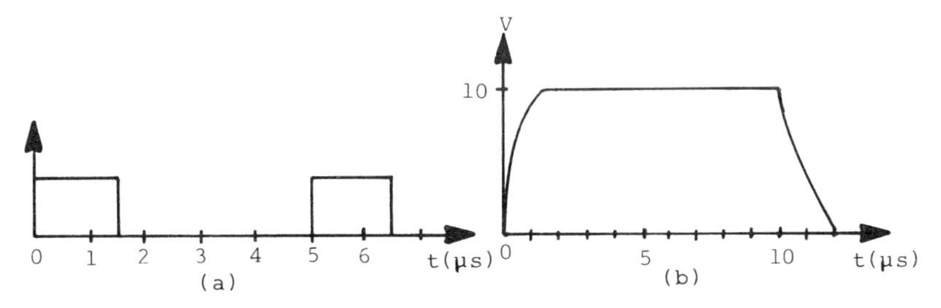

Fig. 12.24 Waveforms for problems 12.1 and 12.2

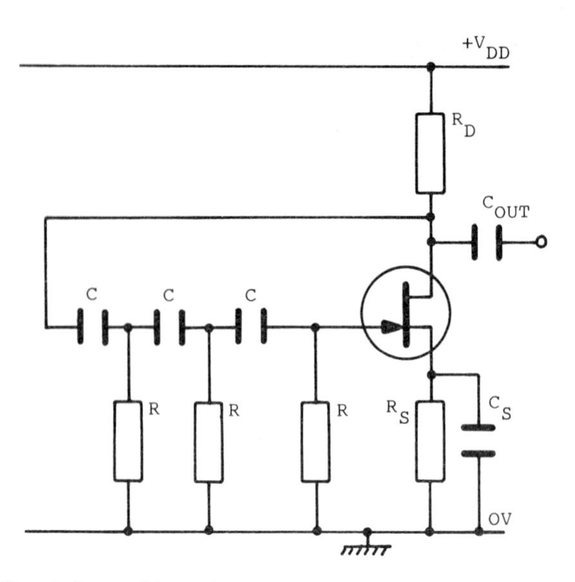

Fig. 12.25 Circuit for problem 12.3

is to be 100 kHz, calculate (a) the required value of the inductor L_1, (b) the minimum frequency of oscillation.

12.5 The circuit of fig. 12.14 has the following component values: $R_1 = R_4 = 2.2$ kΩ, $R_2 = 20$ kΩ, $R_3 = 50$ kΩ, $C_1 = C_2 = 470$ pF. Calculate (a) the frequency of operation, (b) the mark-to-space ratio.

12.6 If the circuit of fig. 12.14 is to have a pulse-repetition frequency of 1 MHz with a mark-to-space ratio of 1:5, calculate the appropriate values for the resistors R_2 and R_3 if $C_1 = C_2 = 470$ pF.

12.7 Explain how an *R–C* network can be used to produce (a) an integrator, (b) a differentiator.

12.8 What are the basic requirements for an oscillator circuit?

12.9 Why are *L–C* oscillators unsuitable for low-frequency applications?

12.10 How is the output of a common-emitter amplifier given a further 180° phase shift in an *L–C* oscillator?

12.11 What are the advantages of using a Wien-bridge oscillator rather than a phase-shift oscillator?

12.12 Define the term 'stability' in relation to oscillators.

12.13 Explain the significant characteristics of the following multivibrators: (a) astable, (b) monostable, (c) bistable.

12.14 What effect will small supply-voltage variations have on the frequency of oscillation of a multivibrator? Give reasons for your answer.

12.15 Why do some applications require the operation of the multivibrator to be synchronised?

12.16 Explain how and why a practical pulse waveform will differ from the ideal.

13 Measurements

13.1 Introduction

It is essential that technicians involved in the testing and service of modern electronic equipment have some understanding of the capabilities and – as important – the limitations of any measurement equipment they may use. While some routine and breakdown servicing does require frequency counters, digital voltmeters, etc., most testing requirements for general electronics can be covered adequately by just two instruments – a good-quality multimeter and a cathode-ray oscillosope (CRO or just 'scope).

For testing complex electronic systems, additional test equipment such as logic analysers or signature analysers may be necessary. In these cases, reference will have to be made to the equipment operating manual. In fact many of the common items of test equipment, such as the multimeter, have capabilities that very few of us would be able to use without referring to the manual.

Which equipment is best for a particular measurement is largely a matter of experience, although equipment manufacturers may specify certain types of meter to be used for particular measurements. The multimeter is suitable for measurement of resistance, most d.c. voltage and current measurements, and low-frequency (particularly 50 Hz mains) a.c. measurements. The CRO can be used to measure d.c. and a.c. voltage and indirectly current, but it is particularly useful where it is necessary to measure phase and/or time relationships. It also has the ability to allow us to 'see' the waveform being examined.

When taking measurements in electronic circuits, it is particularly important that the user is aware of any effects resulting from the connection of the test equipment to the circuit. Suppose we wished to measure the voltage across R_2 in fig. 13.1(a). Inspection should tell us that the p.d. will be 5 V. However, if we used a typical multimeter on the 10 V range we would get a reading of about 1.5 V – apparently a disastrous result, but, because of the effect of the resistance of the meter itself, covered in example 13.2, this is what the meter should read; in fact if it were to indicate 5 V there would be something wrong with either the circuit or the meter.

13.2 Moving-coil instruments

The moving-coil meter consists of a coil of many turns of fine wire mounted at right angles to a magnetic field – usually provided by a permanent magnet – and pivoted so that it is free to rotate. If a current flows in the coil, a deflecting torque will be set up (the motor effect) and,

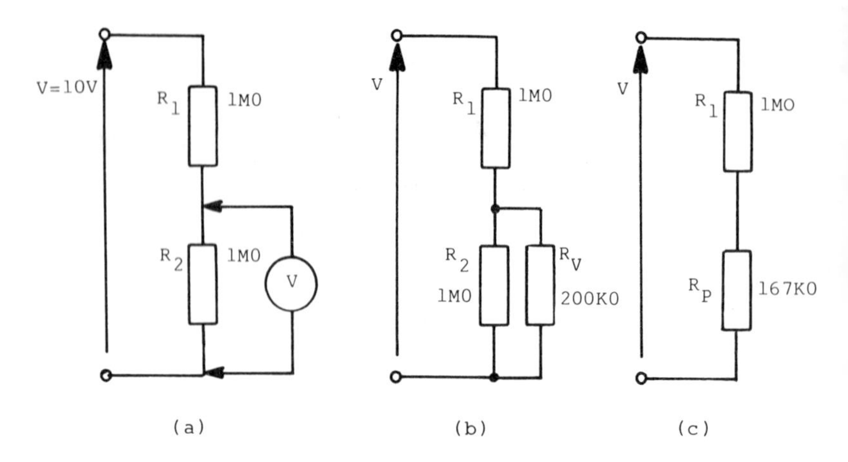

Fig. 13.1 Effect of meter resistance

since the coil is pivoted, it rotates. To prevent the coil from continuing to rotate, hairsprings are used to provide a control torque. The hairsprings are designed to provide a torque which is proportional to deflection, resulting in a linear scale. It is usual to use two contra-wound (in opposition) springs so that their torques cancel out at the zero position. By varying one of these torques, zero adjustment can be carried out. The springs are also used to feed the current into and out of the moving coil. A pointer mounted on the moving-coil shaft gives an indication over a graduated scale. Figure 13.2 shows the basic moving-coil arrangement.

To damp the meter movement, the coil is wound on, but insulated from, an aluminium former. If the former moves in the magnetic field, e.m.f.'s will be induced into it and these e.m.f.'s will produce eddy currents in the former which, according to Lenz's law, will be in such a direction as to oppose the motion producing them, thus damping the oscillations.

Better-quality meters will have jewelled bearings to reduce friction and a mirror placed behind the pointer to reduce parallax errors (errors due to the position of the eyes when taking a reading).

The coil consists of many turns of fine wire, so the current required to achieve deflection is very small (usually less than 100 μA). The moving-coil meter provides the basis for ammeters, voltmeters, and ohmmeters by using suitably connected resistors and/or batteries.

13.3 The voltmeter

A voltmeter is placed in parallel with the circuit or component whose voltage is to be measured and should therefore draw as little current as possible, so it must have a high resistance. Although it draws very little current, the basic moving-coil meter has a very low resistance. To convert the moving-coil meter for use as a voltmeter, it is necessary to add a series resistor, called a 'multiplier'. Example 13.1 shows the method for determining the value of the multiplier.

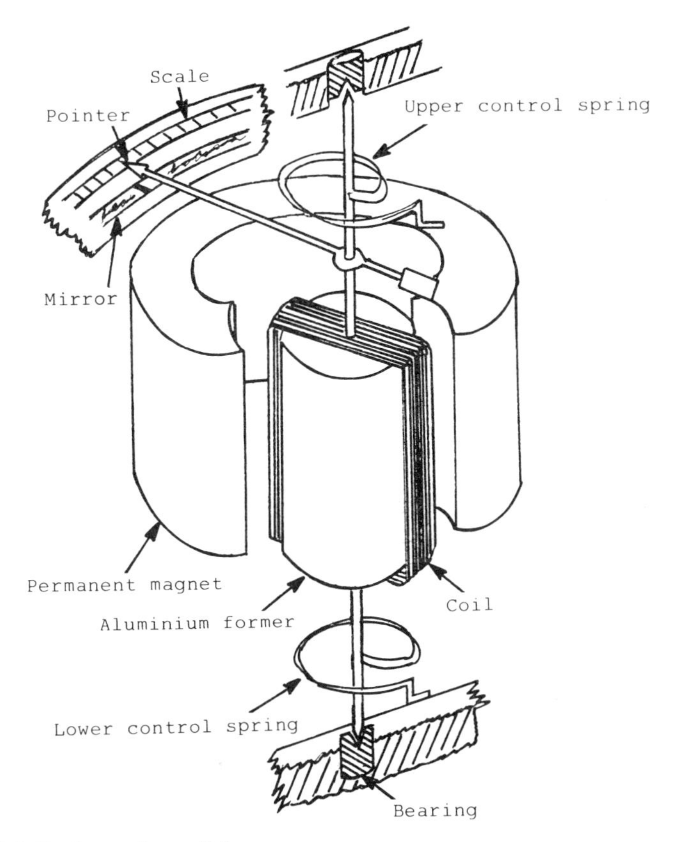

Fig. 13.2 Basic moving-coil instrument

Example 13.1 A permanent-magnet moving-coil meter gives a full-scale deflection (f.s.d.) for a current of 100 μA, and has a resistance of 50 Ω. Show how this can be used as a voltmeter with a range of 0 to 100 V. Find the value of the multiplier required.

Figure 13.3(a) shows the basic circuit connection for the voltmeter.

$$I = \frac{V}{R + R_M} = \frac{V}{R_T}$$

where R_M is the resistance of the meter.

The multiplier R must be chosen so that the current I does not exceed the current required for f.s.d. of the meter.

$$\therefore \quad R_T = R + R_M = \frac{V}{I} = \frac{100 \text{ V}}{100\ \mu\text{A}} = 1 \text{ M}\Omega$$

$$\therefore \quad R = R_T - R_M = 1 \text{ M}\Omega - 50\ \Omega \approx 1 \text{ M}\Omega$$

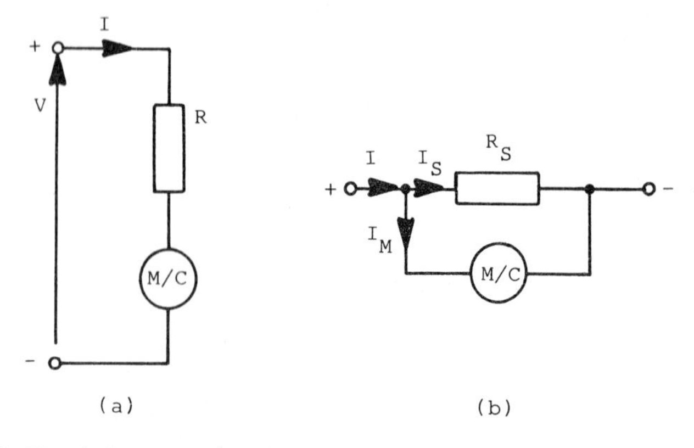

Fig. 13.3 Circuit for examples 13.1, 13.3, and 13.4

In practice the resistance of the meter can be ignored – this is particularly the case if standard resistors are to be used; for example, if a 1 MΩ 1% tolerance resistor were to be used, the tolerance would be ± 10 000 Ω.

13.3.1 Sensitivity of a voltmeter

The 'sensitivity' of a voltmeter is a measure of the current required to give a particular reading. It is given in units of *ohms per volt* and is obtained by dividing the total voltmeter resistance by the voltage required to give f.s.d. For the meter in example 13.1, the sensitivity is 1 MΩ/100 V = 10 kΩ/V. Thus a multimeter with a number of different voltage ranges will have a different resistance on each range.

Example 13.2 A voltmeter having a sensitivity of 20 kΩ/V is to be used on its 10 V range. Calculate

a) the total voltmeter resistance.
b) the actual reading on the meter if it is used to measure the p.d. across resistor R_2 in fig. 13.1(a).

a) Total voltmeter resistance $R_V = 10\text{ V} \times 20\text{ k}\Omega/\text{V} = 200\text{ k}\Omega$

b) As far as the circuit of fig. 13.1(a) is concerned, the voltmeter simply appears to be a resistor of 200 kΩ in parallel with R_2. This is shown in fig. 13.1(b). The parallel combination of R_2 and R_V is

$$R_P = \frac{R_2 \times R_V}{R_2 + R_V} = \frac{1\text{ M}\Omega \times 200\text{ k}\Omega}{1\text{ M}\Omega + 200\text{ k}\Omega}$$

$$= 166.67\text{ k}\Omega$$

$$\approx 167\text{ k}\Omega$$

Figure 13.1(c) shows the equivalent circuit with R_P. The actual voltage which should be indicated by the meter, V_P, is given by

$$V_P = \frac{V_S \times R_P}{R_1 + R_P} = \frac{10\ \text{V} \times 167\ \text{k}\Omega}{1\ \text{M}\Omega + 167\ \text{k}\Omega}$$

$$= 1.43\ \text{V}$$

13.4 The ammeter

The basic moving-coil meter also forms the basis of the ammeter. To enable different current ranges to be measured, it is necessary to extend the range of the basic meter. This is achieved by putting a resistor in parallel with the meter to share the total current. The resistor is called a 'shunt'. By having a range of suitable resistors which can be switched in parallel, various current-measurement ranges can be provided. Example 13.3 shows the method for determining the value of the shunt.

Example 13.3 A permanent-magnet moving-coil meter gives full-scale deflection for a current of 100 μA and has a resistance of 50 Ω. Show how this can be used as an ammeter having a range of 1.0 A.

Figure 13.3(b) shows the circuit diagram. The total current I must be split between the shunt R_S and the meter. Since the current in the meter is proportional to the total current, the meter can be calibrated directly in terms of the total current. The value of the shunt must be chosen so that when $I = 1.0$ A, its maximum value, the meter will draw 100 μA. Thus the current in the shunt is given by

$$I_S = I - I_M = 1\ \text{A} - 100\ \mu\text{A} = 0.9999\ \text{A}$$

Since the shunt and the meter are in parallel, the p.d. across each is the same and is given by

$$V_M = I_M R_M$$

where I_M is the current required to produce full-scale deflection in the meter itself.

$$\therefore\ V_M = 100\ \mu\text{A} \times 50\ \Omega = 5\ \text{mV}$$

The value of the shunt can now be calculated from

$$R_S = \frac{V_M}{I_S} = \frac{V_M}{I - I_M}$$

$$= \frac{5\ \text{mV}}{0.9999\ \text{A}} = 5.0005\ \text{m}\Omega$$

The resistance of the shunt is very small. This means that the shunt will be difficult and expensive to make, particularly since it must be made with very close tolerances if the meter is to be accurate. Also, since the resistance is so small, any 'contact resistance' when assembling the meter will be very important. The larger the current range required, the smaller will be the value of the shunt, providing the limiting factor on the maximum current ranges of most multimeters.

Example 13.4 The coil of a moving-coil instrument has a resistance of 75 Ω, and full-scale deflection is achieved with a current of 1.0 mA. Calculate

a) the value of a suitable resistor to convert the basic meter into (i) a voltmeter with a 0 to 25 V range, (ii) an ammeter with a range 0 to 100 mA;
b) the sensitivity of the voltmeter in (a)(i).

a) (i) Figure 13.3(a) shows the circuit diagram for the voltmeter.

$$\text{Total current } I = \frac{V}{R + R_M} = \frac{V}{R_T}$$

$$\therefore \quad R_T = \frac{V}{I} = \frac{25\text{ V}}{1.0\text{ mA}} = 25\text{ k}\Omega$$

$$\therefore \quad R = R_T - R_M \approx 25\text{ k}\Omega$$

ii) Figure 13.3(b) shows the circuit diagram for the ammeter.

$$I_S = I - I_M = 100\text{ mA} - 1\text{ mA} = 99\text{ mA}$$

$$V_M = I_M R_M = 1\text{ mA} \times 75\ \Omega = 75\text{ mA}$$

$$\therefore \quad R_S = \frac{V_M}{I_S} = \frac{75\text{ mV}}{99\text{ mA}} = 0.7576\ \Omega$$

b) The sensitivity of the meter is given by

$$\frac{25\text{ k}\Omega}{25\text{ V}} = 1\text{ k}\Omega/\text{V}$$

Example 13.5 If the voltmeter of example 13.4 is used to measure the p.d. across R_2 in fig. 13.4(a), calculate

a) the power dissipated by the circuit before the voltmeter is connected;
b) i) the reading on the voltmeter when it is connected,
   ii) the percentage error in the reading;
c) the power dissipated by the voltmeter itself when connected to the circuit.

a) $$\text{Total power } P_T = \frac{V^2}{R} = \frac{(12\text{ V})^2}{30\text{ k}\Omega} = 4.8\text{ mW}$$

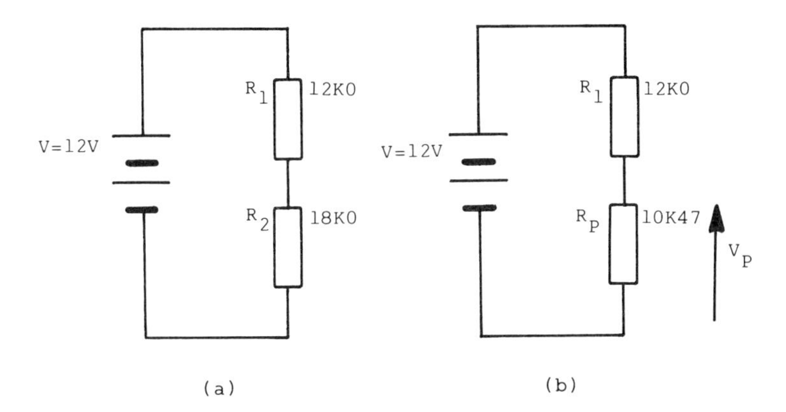

Fig. 13.4 Circuit for example 13.5

b) When the voltmeter is connected, the effective resistance of the voltmeter and R_2 is

$$R_P = \frac{R_2 \times R_M}{R_2 + R_M} = \frac{18\ \text{k}\Omega \times 25\ \text{k}\Omega}{18\ \text{k}\Omega + 25\ \text{k}\Omega} = 10.47\ \text{k}\Omega$$

The equivalent circuit is shown in fig. 13.4(b).

i) The voltmeter reading V_P is given by

$$V_P = \frac{V \times R_P}{R_1 + R_P} = \frac{12\ \text{V} \times 10.47\ \text{k}\Omega}{12\ \text{k}\Omega + 10.47\ \text{k}\Omega} = 5.59\ \text{V}$$

ii) The true reading should be

$$V_{R2} = \frac{V \times R_2}{R_1 + R_2} = \frac{12\ \text{V} \times 10\ \text{k}\Omega}{12\ \text{k}\Omega + 18\ \text{k}\Omega} = 7.2\ \text{V}$$

$$\text{Percentage error} = \frac{\text{indicated value} - \text{actual value}}{\text{indicated value}} \times 100\%$$

$$= \frac{5.59\ \text{V} - 7.2\ \text{V}}{5.59} \times 100\%$$

$$= -28.8\%$$

c) Power dissipated by the voltmeter P_M

$$= \frac{V_P{}^2}{R_M}$$

$$= \frac{(5.59\ \text{V})^2}{25\ \text{k}\Omega}$$

$$= 1.25\ \text{mW}$$

13.5 The ohmmeter

The simple ohmmeter consists of a moving-coil meter, a battery, a fixed resistor R_1, and a variable resistor R_2 as shown in fig. 13.5.

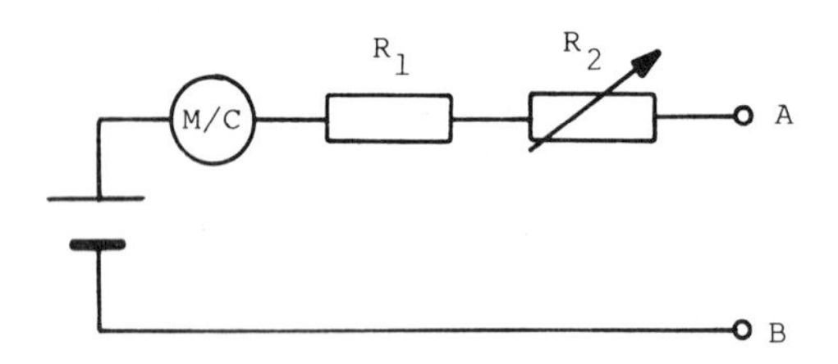

Fig. 13.5 Simple ohmmeter

The fixed resistor is a limiting resistor to ensure that the current through the moving coil does not exceed that for full-scale deflection when the variable resistor is at a minimum. Before use, the terminals A and B must be shorted together to set up, or 'zero', the meter. Resistor R_2 is adjusted until the meter is at full-scale deflection – the zero-ohm position. If an unknown resistor R is connected across the terminals, the current will depend on the total resistance R_T. If R is large the current is small, and if R is small the current is large. The meter is calibrated directly in ohms. The scale is non-linear, being cramped at the high-resistance (low-current) end of the scale. Range extension can be achieved by switching suitable shunts across the meter movement.

These meters are not suitable for very accurate resistance measurements or for measuring insulation resistances, which require a high voltage to be applied to make the measurement.

13.6 The Wheatstone bridge

The Wheatstone bridge provides a simple comparison method for obtaining accurate resistance measurements up to about 100 kΩ, the accuracy depending on the value of the standard resistors in the 'bridge'.

Figure 13.6 shows the basic circuit of the Wheatstone bridge. The resistors R_1, R_2, R_3, and R_X are the 'arms' of the bridge. In practice, any one of these resistors could be the unknown resistor to be measured – we will assume it to be R_X. The resistors R_1, R_2, and R_3 will be standard resistors of known value. To measure resistance, the values of one or more of the standard resistors are adjusted until 'balance' occurs. At balance, the potentials at points 'A' and 'B' are the same – this is detected by the meter connected across the bridge. The detector is generally a centre-zero galvanometer, which is a very sensitive moving-coil meter. The values of the resistors are adjusted until no deflection occurs on the meter; hence no current then flows in the detector, and the potentials at points A and B must be the same. The bridge is then said to be 'balanced'. This is a 'null'

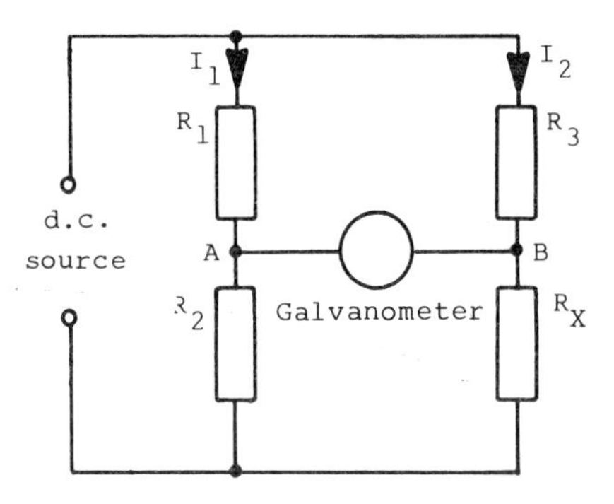

Fig. 13.6 Basic Wheatstone-bridge circuit

method of measurement – the accuracy does not depend upon the meter, just on the standard resistors.

At balance, the current I_1 will flow in R_1 and R_2 and the current I_2 will flow in R_3 and R_X.

$$\therefore \quad I_1R_1 = I_2R_3 \qquad 13.1$$

$$\text{and} \quad I_1R_2 = I_2R_X \qquad 13.2$$

Dividing equation 13.1 by equation 13.2 gives

$$\frac{I_1R_1}{I_1R_2} = \frac{I_2R_3}{I_2R_X}$$

$$\therefore \quad \frac{R_1}{R_2} = \frac{R_3}{R_X}$$

Thus, if R_X is the 'unknown' resistor and R_3 is variable, R_X can be found from

$$R_X = \frac{R_2}{R_1} \times R_3 \qquad 13.3$$

In equation 13.3 it is only necessary to know the value of R_3 and the *ratio* R_2/R_1. Because of this, the resistors R_1 and R_2 are often called the 'ratio arms' and, for ease of calculation, in commercial bridges they are generally made to give convenient ratios such as 0.01, 0.1, 1, 10, etc.

Example 13.6 In fig. 13.6, if balance is achieved with $R_1 = 100\ \Omega$, $R_2 = 1\ \text{k}\Omega$, and $R_3 = 154\ \Omega$, calculate the value of the unknown resistor R_X.

Using equation 13.3,

$$R_X = \frac{R_2}{R_1} \times R_3 = \frac{1000}{100} \times 154\ \Omega = 1540\ \Omega$$

13.7 Rectifier instruments

The basic moving-coil instrument is not suitable for use on a.c., since an oscillating magnetic field would simply cause the pointer to vibrate about the zero position. The moving-coil meter can be converted for use on a.c. with a rectifier system. It is generally necessary to use a transformer in conjunction with the rectifier. Figure 13.7(a) shows the basic arrangement. The input impedance of the meter is now that of the transformer, and this is much lower than the d.c. resistance. A typical sensitivity is 1000 Ω/V.

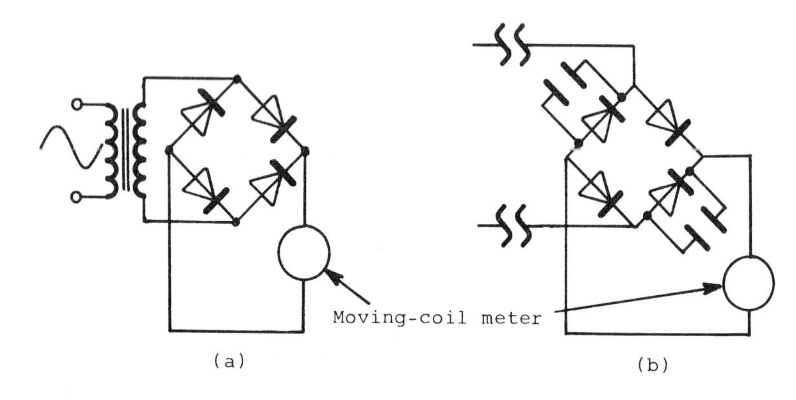

Fig. 13.7 (a) Simple rectifier instrument (b) half-cycle equivalent circuit of fig. 13.7(a)

Rectifier instruments produce a deflection which is proportional to the *average* power in the waveform but are calibrated to read the r.m.s. value of the waveform. This restricts them to a particular waveform – usually a sine wave – use on other waveforms will introduce a form-factor error. (The form factor is the ratio of the r.m.s. value to the average value of an alternating waveform and is 1.1 for a sine wave.)

Rectifier instruments are also limited in the range of frequencies over which they are satisfactory – up to about 10 kHz. This is due to the bridge rectifier system. Two of the diodes are reverse-biased on each half cycle. Because a reverse-biased p–n junction consists of a high-resistance depletion layer separated by two low-resistance areas effectively forming a capacitor, this effective capacitor is in parallel with the meter. At low frequencies the reactance of this effective capacitor is high and its effects are negligible; however, as the frequency increases, the reactance falls – short-circuiting the meter. Figure 13.7(b) shows the equivalent circuit for one half cycle.

Since the diode characteristics are non-linear at small values of voltage, errors are also introduced if the voltage to be measured is small compared to the forward volt drop across the diode. In this case germanium diodes would be more suitable.

13.8 Cathode ray oscilloscope (CRO)

The CRO is probably the most versatile instrument available for use with modern electronic circuits, if only for its ability to allow us to 'see' waveforms.

13.8.1 The cathode-ray tube (CRT)

The 'heart' of all CRO's is the cathode-ray tube, fig. 13.8. The CRT consists of three functional areas – the electron gun, the focussing or electron-lens system, and the deflection system – and the display screen, all contained in a glass tube. The tube has all the air removed and it is sealed to maintain the vacuum inside.

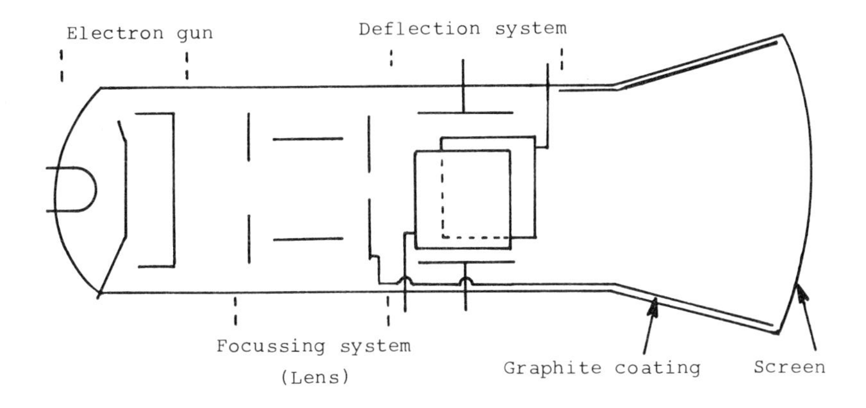

Fig. 13.8 Basic cathode-ray tube

The electron gun produces electrons for the 'beam' by thermionic emission from the heated cathode. (Thermionic emission is achieved when the temperature of a material is raised high enough to give some electrons sufficient thermal energy to allow them to 'escape' from the material if a suitable electric field is applied.) A control grid consisting of a cylindrical container with a hole in the end to allow electrons to pass through covers the cathode. Making the potential on the control grid negative with respect to the cathode concentrates the electrons into the beam and also determines the number of electrons in it, and thus the brightness.

The focussing system consists of three plates, or anodes, held at high positive potentials with respect to the cathode, fig. 13.9. The third anode is at a higher potential than the first, which is at a higher potential than the second. The high positive potential of the first anode attracts the electrons from the cathode and accelerates them towards the screen, passing them through a hole in the first anode on the way. Due to mutual repulsion between them, the electrons in the beam will tend to spread out, producing a fuzzy image on the screen. By controlling the potential on the second

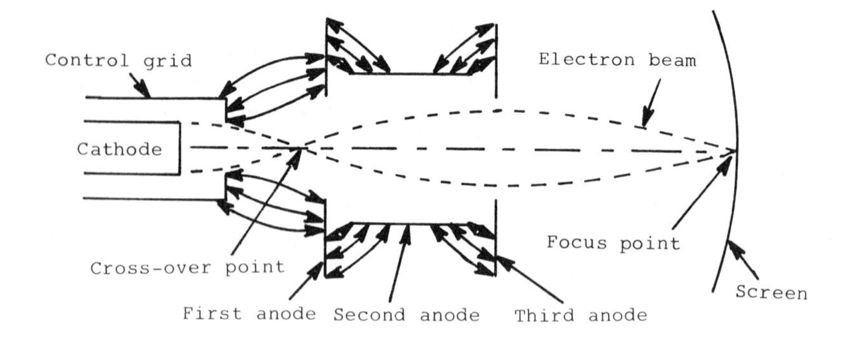

Fig. 13.9 CRT focussing system

anode, the resulting electrostatic field can be used to focus the beam. The field between the grid and the first anode causes the beam to 'cross over' just before it enters the anode system. As an electron tends to move from a lower to a higher potential, any electrons moving away from the axis will be deflected towards the axis and then away from it. As the electron is accelerating all the time, assisted by the potential on the third anode, it is less affected by the diverging field than by the converging one, so the beam is brought to a focus on the screen.

The screen consists of a coating of a fluorescent material on the inside of the tube face. The degree of brilliance and afterglow can be controlled by suitable choice of material. Due to the high velocity of the electrons before impact, some secondary electrons may be emitted from the screen. These are collected by a conductive graphite coating, which is connected to the final anode, on the inside of the tube.

To do anything useful with the beam which has been focussed on the screen, it is necessary to be able to move or deflect it. Deflection is achieved by using two pairs of plates arranged as shown in fig. 13.10. These plates are at about the potential of the final anode, so as not to affect the speed of the electrons. By adjusting the p.d. between each pair of plates, the field produced will deflect the beam. The plates producing vertical deflection are called the *Y*-plates and those producing horizontal deflection are called the *X*-plates.

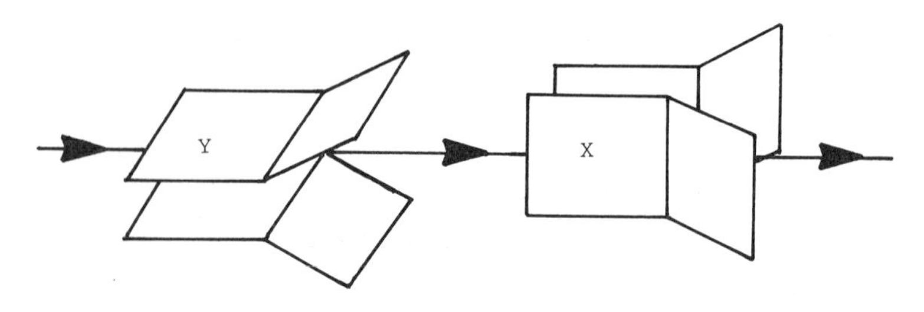

Fig. 13.10 CRT deflection system

Many of the applications in modern electronics require the use of two beams, e.g. when comparing the phase relationship between the input and output signals of an amplifier. Early oscilloscopes achieved this by using a 'beam-splitter' plate between the Y-plates, but this method is no longer used. Almost all modern oscilloscopes now use a 'beam-switching' method. Basically, the single beam is shared between two traces by using a fast switching system. There are two basic methods of achieving this:

i) *Alternate traces* One trace is displayed on each alternate traverse of the beam across the screen and, because of the persistence of vision, both traces are visible simultaneously.
ii) *Chopped traces* The beam is switched at a very high rate between each trace. This occurs many times in each traverse, giving the impression of a continuous beam.

13.9 The cathode-ray tube as an indicating device

A signal voltage applied to the Y-plates of a CRT would simply produce a vertical trace on the screen. To observe the waveform of a signal, it is necessary to move the spot across the screen at a steady rate. This is done by applying a steadily increasing voltage to the X-plates. Once the spot has been fully deflected, it must be returned to the start position as quickly as possible; thus the deflecting voltage must be reduced as quickly as possible. In practice this takes a finite time, called the 'flyback' time. This system is called a 'time-base voltage'. Figure 13.11 shows a typical waveform. When it is applied to the X-plates it will cause the spot to sweep across the screen, fly back rapidly, and repeat the process over and over again.

To enable the CRO to measure voltages of different amplitudes and frequencies, some means of controlling the signals applied to the X- and Y-plates is required.

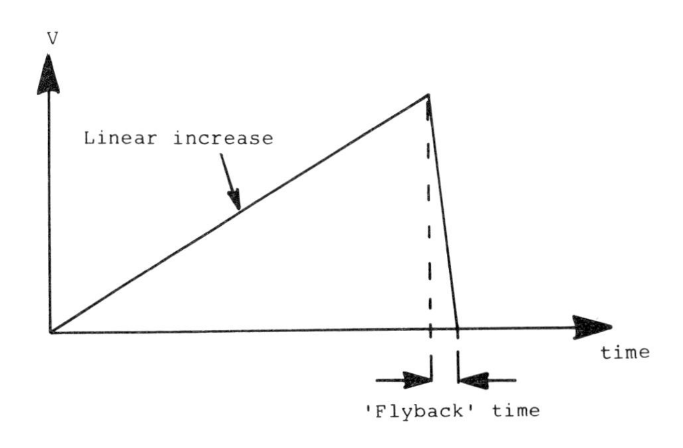

Fig. 13.11 Time-base waveform

The Y direction of the amplifier is used to plot the signal voltage. In order to allow the CRO to display signals from a few millivolts to tens of volts, amplifiers are built into the instrument and are fitted with a gain control. For operating convenience, this is a switched control calibrated directly in volts/cm. If this control is set to 2 volts/cm and the display is 3 cm high, then the actual signal voltage is 2 V/cm × 3 cm = 6 V. Some CRO's also have a variable gain control in addition to the switched one. This should normally be set in its calibrated ('CAL') position. If two traces are available, each trace will have its own Y or vertical amplifier.

The time-base or X-deflection circuit which drives the spot across the screen at a controlled rate also has a switched control calibrated directly in ms/cm, μs/cm, or ns/cm. Again there may be a variable control allowing intermediate speeds to be obtained – this control should be set to 'CAL' for normal use.

13.10 Applications of the CRO

It is not the function of this textbook to give comprehensive instructions on the use of the CRO – this is best achieved in practical work using the equipment handbook for the particular instrument, or one of the many excellent textbooks on the use of CRO's. We shall simply look at some of the basic measurement techniques, assuming that such fundamental requirements as the brightness, focussing, and positioning of the spot etc. are correct.

13.10.1 Voltage measurement

Both d.c. and a.c. voltages may be measured. Most modern CRO's have an AC/DC input switch which in the AC position 'blocks' any d.c. component in the input voltage. This permits small a.c. components which are superimposed on d.c. levels to be examined, e.g. the 'ripple' on rectified mains. When measuring a.c. voltages, the time-base may be used to show the shape of the waveform but it is not essential, since we are only interested in the vertical height of the waveform.

Method

a) Set up the circuit as shown in fig. 13.12(a).
b) With the time-base off, the vertical height of the waveform can be measured. The Y-amplifier control should be adjusted to give the largest trace possible.
c) i) For d.c., the voltage is given by

$$V = y \text{ cm} \times Y\text{-amplifier setting (V/cm)}$$

ii) For a.c., the peak-to-peak voltage is given by

$$V_{\text{P-P}} = y \text{ cm} \times Y\text{-amplifier setting (V/cm)}$$

$$V_{\text{RMS}} = \frac{V_{\text{P-P}}}{2} \times 0.707$$

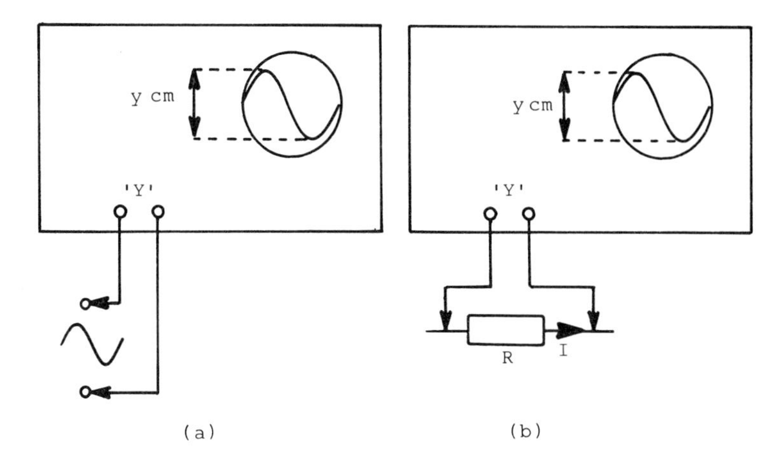

Fig. 13.12 CRO measurements: (a) voltage, (b) current

The CRO has a relatively high input impedance, typically 1 MΩ, which is constant over a wide range of frequencies. When measuring voltages in high-impedance circuits, it may be necessary to increase the effective impedance of the CRO by using a 'probe'.

13.10.2 Current measurement

Current measurement cannot be done directly with a CRO but requires a resistor of known value – a low value such as 1 Ω is ideal. The known resistance must be connected in series with the current to be measured.

Method

a) Set up the circuit as shown in fig. 13.12(b).
b) Measure the voltage across R as in the previous section.
c) Calculate the current using Ohm's law. If a 1 Ω resistor is used, the current will be numerically equal to the voltage, giving a direct reading in amperes.

i) d.c. $I = \dfrac{y \text{ cm} \times \text{V/cm}}{R}$

ii) a.c. $I = \dfrac{y \text{ cm} \times \text{V/cm}}{2R} \times 0.707$

13.10.3 Frequency measurement

a) Set up the circuit as shown in fig. 13.13(a).
b) Ensure that the time-base variable control is set to 'CAL'. Adjust the frequency of the time-base control until the smallest number of complete cycles is obtained on the screen, if possible just one.

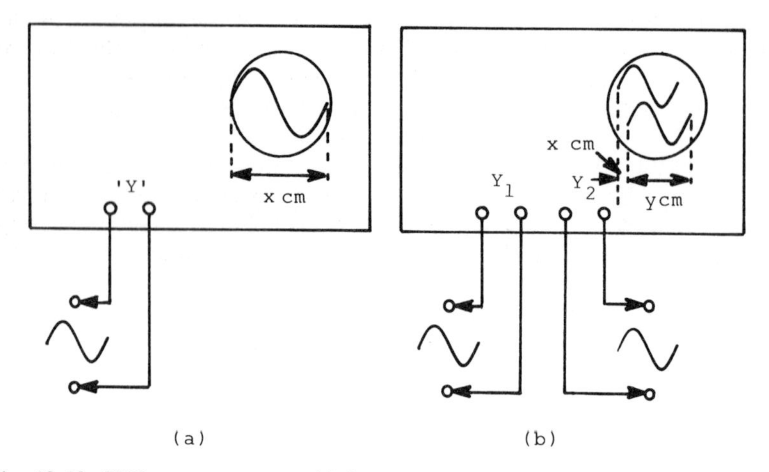

Fig. 13.13 CRO measurements: (a) frequency, (b) phase difference

c) Measure the 'length' of one complete cycle. Calculate the periodic time T from

$$T = x \text{ cm} \times \text{sensitivity (s/cm)}$$

Hence calculate $f = 1/T$.

13.10.4 Measurement of phase difference

a) Set up the circuit as shown in fig. 13.13(b).
b) Ensure that the time-base variable control is set to 'CAL'. Adjust the frequency of the time-base control until the smallest number of complete cycles is obtained on the screen, if possible just one of each waveform.
c) Measure the lengths x and y. Phase difference ϕ can be calculated from

$$\phi = \frac{x}{y} \times 360°$$

Frequency and phase difference can also be measured using a single-beam CRO and 'Lissajou's figures'. However, the method requires an external frequency source, such as a signal generator. The external signal is used to operate the CRO time-base. The waveform whose frequency is to be measured is applied to the Y-input, and the comparison frequency is adjusted until a pattern is formed on the screen.

The shape of the figure produced will depend on the frequency and phase difference of the two waveforms. If the two frequencies are identical and in phase, a straight line will be produced. If the phase difference is 90°, a circle is produced. Intermediate values of phase difference produce an ellipse. (Note: a circle will be produced only if the amplitude of the waves is the same and the X- and Y-amplifier sensitivities are the same or adjustment is made to either the X- or Y-sensitivity controls). Multiples and submultiples of the comparison frequency also produce standard patterns. The method is not very convenient to use.

The major advantage of this method is that the accuracy is not dependent on the CRO, only on the comparison frequency. Where critical adjustment of frequency is required, a standard frequency generated using a crystal-controlled oscillator may be used in conjunction with the method.

In general, a technician involved in servicing will almost certainly have a double-beam CRO available which should be adequate for the majority of applications.

Example 13.7 A cathode-ray-tube display shows one complete cycle of a sine wave which has a total height of 8 cm and a width of 10 cm. If the CRO Y-amplifier setting is at 5 V/cm and the time-base setting is 200 μs/cm, calculate

a) the r.m.s. voltage of the applied waveform,
b) the frequency of the waveform.

The waveform is shown in fig. 13.14(a).

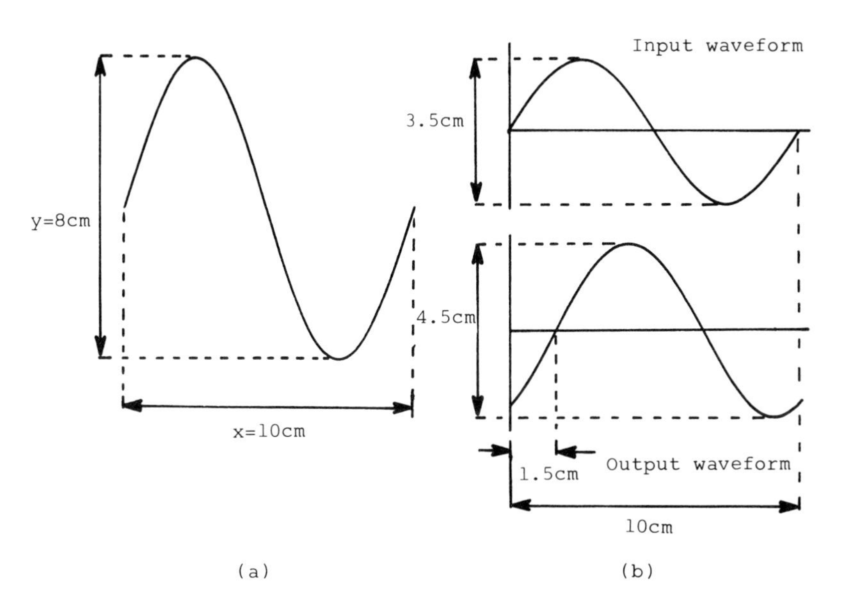

Fig. 13.14 Waveforms for example 13.7 and problem 13.8

a) $V_{RMS} = \dfrac{V_{P-P} \times 0.707}{2}$

$= \dfrac{Y\text{-sensitivity} \times y \text{ cm} \times 0.707}{2}$

$= \dfrac{5 \text{ V/cm} \times 8 \text{ cm} \times 0.707}{2}$

$= 14.14$ V

b) Periodic time $T = X$-sensitivity $\times\ x$ cm

$= 200\ \mu\text{s/cm} \times 10$ cm

$= 2000\ \mu$s

$= 2$ ms

Frequency $f = 1/T$

$= 1/(2 \text{ ms})$

$= 500$ Hz

Problems

13.1 A moving-coil instrument has a resistance of 2.76 kΩ and requires a current of 37.5 μA to give full-scale deflection.

a) Show how the instrument can be converted into an ammeter having a range higher than 37.5 μA and calculate the value of any additional component(s) required to convert the basic meter to an ammeter with a range of (i) 0 to 50 μA, (ii) 0 to 1 A.

b) Show how the instrument can be converted into voltmeter with a range of (i) 0 to 3 V, (ii) 0 to 10 V.

c) What voltage range could the meter be used on without any additional components?

13.2 Two resistors, of 270 kΩ and 560 kΩ respectively, are connected in series across a 9 V d.c. supply. A voltmeter with a sensitivity of 20 kΩ/V is to be used to measure the p.d. across the larger one. Calculate

a) the actual p.d. across the 560 kΩ resistor;

b) the meter reading when used on (i) the 10 V range, (ii) the 100 V range.

13.3 a) Calculate the current in the resistor of fig. 13.15(a).

b) An ammeter on its 50 μA range which has an internal resistance of 2 kΩ is to be used to measure the current in the 180 kΩ resistor as shown in fig. 13.15(b). Calculate the meter reading.

c) A voltmeter with a sensitivity of 20 kΩ/V is connected as shown in fig. 13.15(c). Calculate the meter reading if it is used on its 10 V range.

d) What will the voltmeter in (c) read if it is connected across the supply terminals?

13.4 A voltmeter of sensitivity 20 kΩ/V is to be used to measure the base voltage in the circuit of fig. 13.16. Calculate

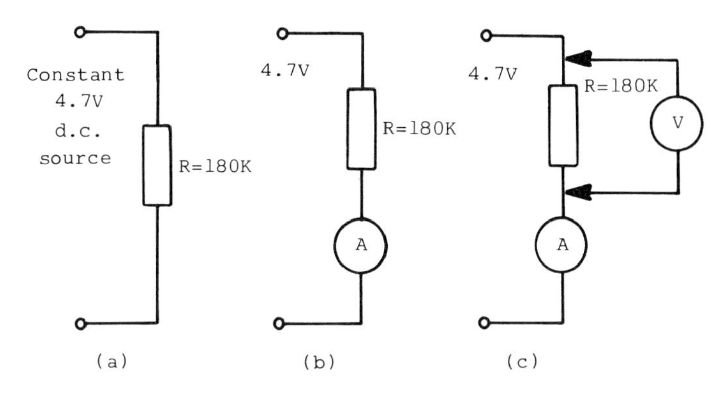

Fig. 13.15 Circuit for problem 13.3

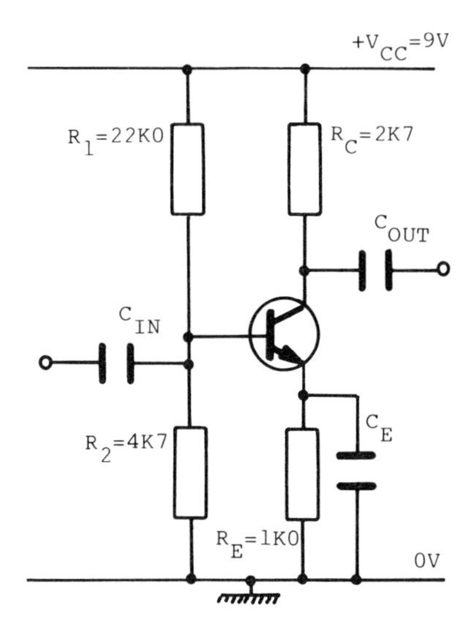

Fig. 13.16 Circuit for problem 13.4

a) the actual base voltage if the transistor input resistance can be considered to be 250 kΩ;
b) i) the voltmeter reading on the 3 V range,
   ii) the percentage error in the reading in (b)(i);
c) i) the voltmeter reading on the 10 V range,
   ii) the percentage error in the reading in (c)(i).

13.5 The Wheatstone bridge in fig. 13.17 is used to measure the value of the unknown resistor R_X. Balance is achieved with $R_1 = 10$ kΩ, $R_2 = 1$ kΩ, and $R_3 = 1476$ Ω.

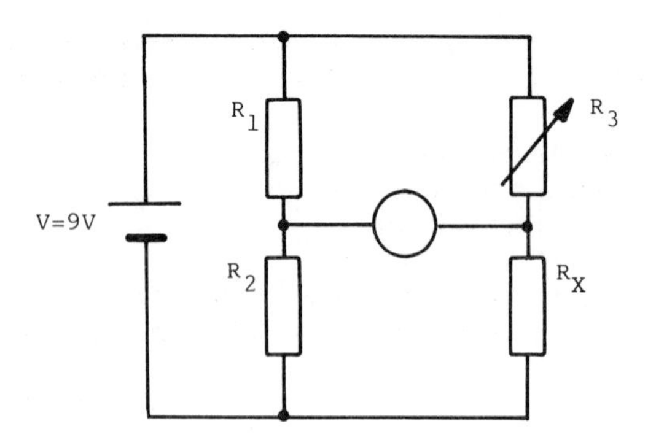

Fig. 13.17 Circuit for problems 13.5 and 13.6

a) Calculate the value of R_X.
b) Calculate the maximum and minimum values between which R_X lies if the tolerance for each of the three resistors in the bridge is $\pm 0.1\%$.
c) If the battery potential falls to 8.5 V, what effect will this have on the accuracy of any reading?

13.6 Balance is obtained in the circuit of fig. 13.17 with $R_3 = 75\ \Omega$ and $R_X = 150\ \Omega$. When R_X is replaced with an unknown resistor, balance is then obtained with $R_3 = 175\ \Omega$. Calculate the value of the unknown resistor.

13.7 Figure 13.18 shows the waveform displayed on a CRO. If the time-base setting is 5 μs/cm and the Y-amplifier sensitivity is 20 mV/cm, calculate

a) the amplitude of the waveform,
b) the repetition frequency of the waveform.

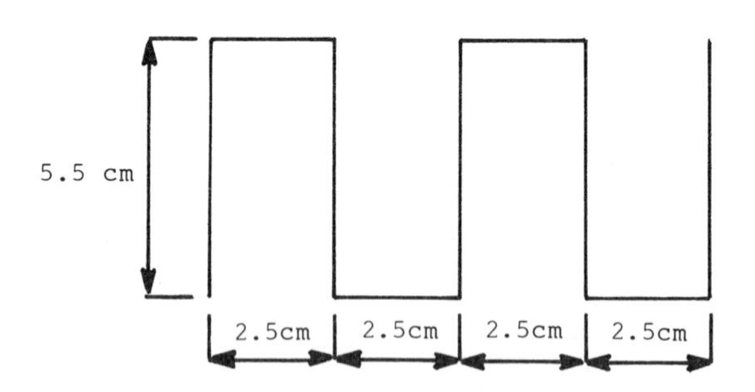

Fig. 13.18 Waveform for problem 13.7

13.8 A double-beam CRO is used to observe the input and output waveforms of a common-emitter amplifier. The display produced is shown in fig. 13.14(b). The input waveform is displayed on the Y_1 trace with an amplifier sensitivity of 50 mV/cm; the output is displayed on the Y_2 trace with an amplifier sensitivity of 2 V/cm. The time-base setting is 0.2 μs/cm. Calculate

a) the r.m.s. value of the input voltage,
b) the r.m.s. value of the output voltage,
c) the voltage gain of the amplifier,
d) the frequency of the input signal,
e) the phase lag between the input and output waveforms.

13.9 Why is it difficult to produce multimeters for very low current measurements?

13.10 What are the advantages of using a 'bridge' system to measure resistance?

13.11 Explain why the basic moving-coil instrument is not suitable for use on a.c.

13.12 Why is the sensitivity of a moving-coil multimeter for use on a.c. much less than that when it is used on d.c.?

13.13 What is the main factor limiting the upper frequency at which a moving-coil multimeter may be used?

13.14 Why are moving-coil multimeters likely to give large errors when used to read small a.c. voltages?

13.15 What are the advantages of using a CRO for taking measurements in a.c. circuits?

13.16 A typical input resistance of a CRO, without a probe, is 1 MΩ. A multimeter with a sensitivity of 20 kΩ/V on the 100 V range has an input resistance of 2 MΩ. Explain why the CRO is more likely to give accurate readings over a range of voltages.

Appendix: practical work

Introduction

Hopefully, any student following a technician course resulting in the award of a certificate or diploma will either be working in electronics or be following a course at a technical college in which practical work is an essential element of the course. In the majority of cases the practical work will have been determined by the lecturers teaching the subject material.

The practical work and exercises in this appendix are not meant to be exhaustive or the only means of covering the practical work involved in the TEC units.

Practical work can be used to achieve a number of objectives for both teacher and student. Initially, most students do not have experience in the use of simple test and measurement instruments such as multimeters and oscilloscopes (a point often ignored by lecturers), so any practical work should include familiarisation with the relevant equipment as an integral part of any measurement exercise – if necessary, insisting on the use of the equipment handbook. To achieve practice in using meters and to build up the student's confidence, interpret simple circuit diagrams – again, it is often assumed that the student can read a circuit diagram when in fact this is a new 'language' which the majority of us must learn first. Simple exercises such as the measurement of diode and transistor characteristics can be used for this purpose. If the objective is purely to show the shape of the characteristics, it is much easier to use data books. Finally, practical work can be used to demonstrate or verify theory.

Equipment

Practical work on electronic equipment inevitably requires the use of instruments. The absolute minimum is a multimeter, an oscilloscope, a power supply, a signal generator, and some means of constructing the circuits.

The circuits required for the student to use to achieve the practical objectives can be available on permanently assembled boards. However, this approach is very limited because of its 'dedication', and it is wasteful of resources. All the practical exercises following were built and tested using the Locktronics boards and components. This system is extremely versatile, enabling the student to construct relatively complex circuits rapidly and, as importantly, permitting the lecturer to quickly locate faulty connections. Figure A1.1 shows a typical two-stage common-emitter circuit layout using a Locktronics kit. If the standard components provided in the Locktonics kits are not felt to be suitable, then it is a simple

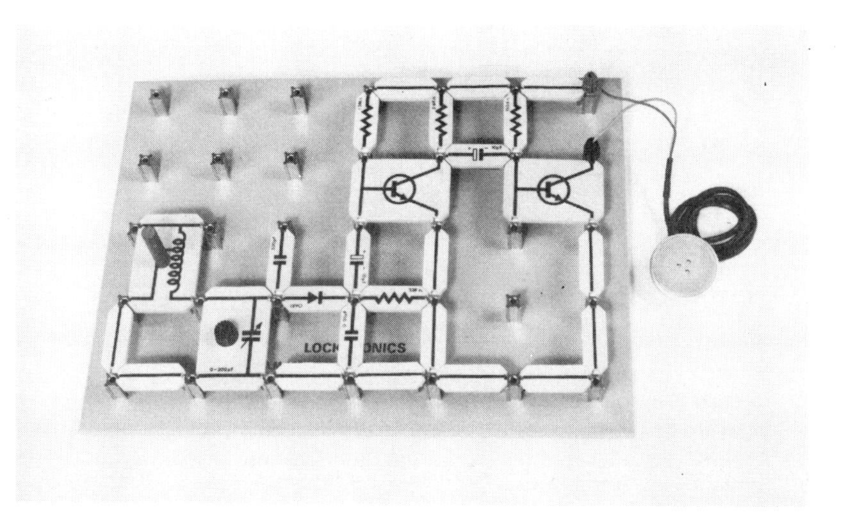

Fig. A1.1 Locktronics board layout

matter to construct tailor-made components using the blank pieces available – the method used for the circuits in this book. The exercises can, however, be easily adapted for use with the available Locktronics components.

The actual equipment used for the verification was

a) multimeter – Avometer model 8 mark 5;
b) oscilloscope – Farnell double-beam DTV12-14;
c) signal generator – Feedback function generator FG600;
d) digital multimeter – Fluke 8021B;
e) Locktronics LK750 baseboard and components;
f) power supply (home-made) — 0 to 18 V, 500 mA. (A power supply with current-limit capability would be an advantage for many of the characteristic measurements.)

When taking measurements, always use the meter on the highest range available until you have established the maximum reading which will be obtained.

Exercise A1.1 Diode measurements

1. Connect up the circuit of fig. A1.2 using a germanium diode, e.g. an OA47. Ensure that the potentiometer is set to minimum.
2. From the data sheets, find the maximum current which the diode can be allowed to carry.
3. Adjust the potentiometer to increase the voltage from zero until current flow is just detectable and note the voltmeter reading. What is the significance of this reading?

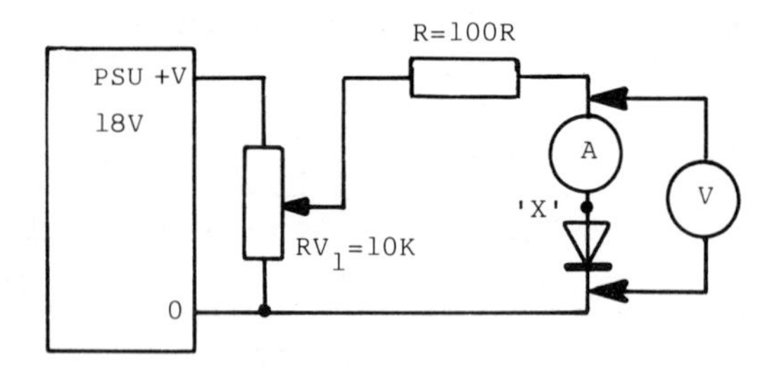

Fig. A1.2 Circuit for exercise A1.1

4. Record the voltmeter and ammeter readings for a number of further steps up to the maximum allowable diode current.
5. Reverse the diode and record the current flow for the maximum supply voltage available.
6. Repeat steps 3 and 4 with the voltmeter connected to point 'X'. Comment on any differences in the readings.
7. Repeat steps 1 to 5 using a silicon diode such as a 1N4001. What differences do you observe? Do they agree with theory?
8. If time and a digital voltmeter are available, repeat step 6. Can you explain any differences?
9. Plot the characteristics for both diodes on the same graph.

Exercise A1.2 Zener-diode measurements

1. Connect up the circuit of fig. A1.3 using a BZY88C4V7 zener diode.
2. *Calculate* the maximum permissible current for the zener diode. (You need to find the power rating of the diode.)
   I_{DMAX} = ...
3. If the power supply has current-limit facility, set this to the current value found in step 2. If this is not available, use resistor $R_X = (V_{MAX} - V_D)/I_{DMAX}$.
   Adjust the potentiometer in 0.5 V steps until the voltage starts to stabilise – i.e. very small voltage changes produce relatively large

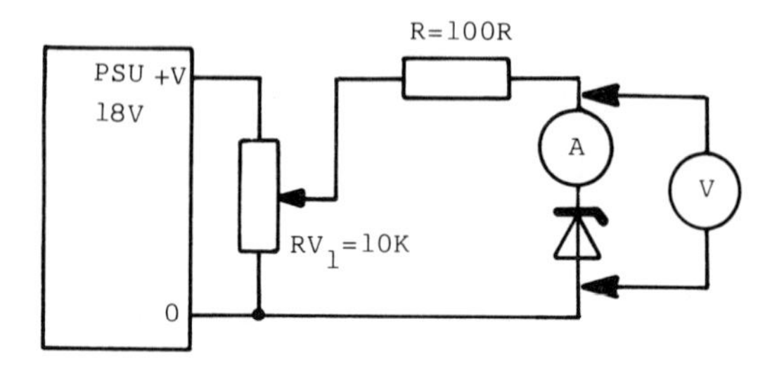

Fig. A1.3 Circuit for exercise A1.2

current changes – recording the value of diode current. (Note: you may have to use the low-current ranges of the ammeter for some of the initial readings, so take care to change ranges when you are near full-scale deflection on one range.)

Once the zener has stabilised, record the small voltage changes for convenient steps of current up to the maximum permissible for the diode.

4. Reverse the zener diode and obtain the forward characteristics.
5. Plot the forward and reverse characteristics on the same graph.

   Observe the characteristics and note that the forward characteristics are identical to those of an ordinary silicon diode – it is after all only a silicon diode.

Exercise A1.3 Common-base and common-emitter characteristics

The circuits of figs A1.4 and A1.5 are typical circuits for measuring common-base and common-emitter characteristics. Minor changes in component values may be required to accommodate different transistor types.

Note: when measuring two variables while holding a third constant, such as measuring I_E for changes in V_{BE} with V_{CB} held constant, it will generally be necessary to continually check and if required adjust the 'constant'.

a) Common-base characteristics for a BC109 transistor

1. Connect up the circuit of fig. A1.4.
2. Set V_{CB} to 5 V.
3. By varying V_{BE}, obtain values for I_E up to 10 mA. (This is well below the maximum possible for a BC109.)
4. Plot the common-base input characteristics.
5. Set I_E to 2 mA. For convenient (*you* decide what is convenient) values of V_{CB}, obtain the corresponding values of collector current I_C.

   Repeat for values of I_E of 4 mA, 6 mA, 8 mA, and 10 mA.
6. Plot the output characteristics for the data obtained in step 5 and obtain values of h_{ob} for $I_E = 6$ mA and of h_{fb} for $V_{CB} = 6$ V.

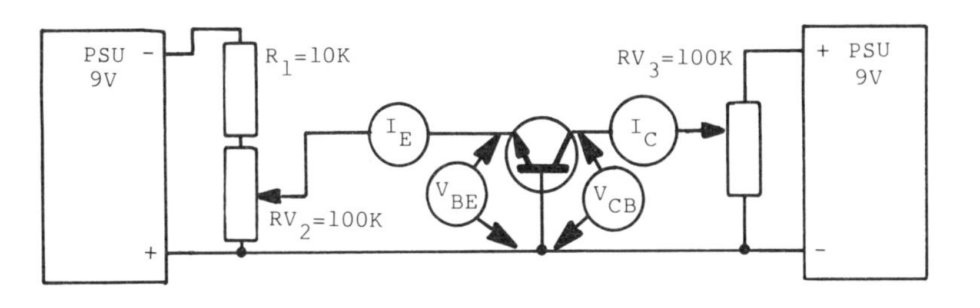

Fig. A1.4 Circuit for exercise A1.3 – common-base

b) Common-emitter characteristics for a BC109 transistor

1. Connect up the circuit of fig. A1.5.
2. Set V_{CE} to 5 V.
3. Determine experimentally the value of I_B required to make I_E = 10 mA, then obtain values of V_{BE} and I_B for I_B up to this value.
4. Choose five values of I_B from 0 μA to the maximum obtained in step 3. Set I_B constant at each of these values in turn and vary V_{CE} from 0 V to 10 V in convenient steps to obtain values for I_C.
5. Plot the input and output characteristics for the data obtained above. From the appropriate graph, determine values for

   a) h_{ie},
   b) h_{FE} for V_{CE} = 5 V,
   c) h_{oe} for I_B at some convenient value.

   Compare the values obtained with typical values which can be obtained from the data sheets at the end of chapter 2.

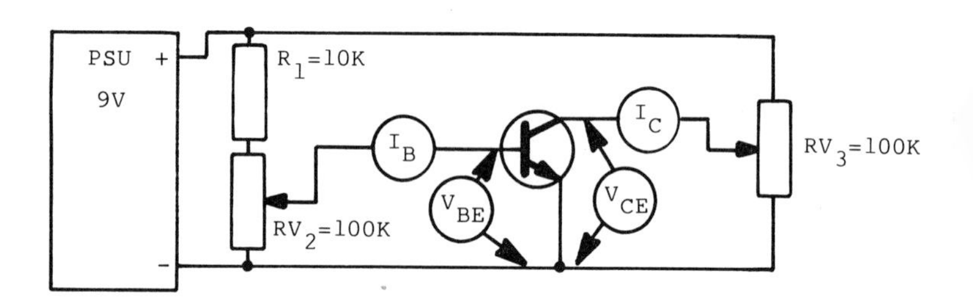

Fig. A1.5 Circuit for exercise A1.3 – common-emitter

Exercise A1.4 Simple transistor tests

If we look closely at the physical construction of the transistor as shown in fig. 2.1 in chapter 2, we can see that the transistor can be represented as two diodes 'back-to-back', as shown in fig. A1.6. (Note: this is a theoretical representation and a transistor cannot be constructed this way.) This simple equivalent circuit enables us to carry out a number of simple tests on a transistor.

If we use a simple ohmmeter to measure the resistance between the transistor terminals, the following results will be produced:

a) between base and emitter a 'high' resistance and a 'low' resistance, depending on which way round the terminals of the meter are connected;
b) between base and collector a similar pair of readings to those obtained in (a) above;
c) between collector and emitter a 'high' resistance whichever way the meter leads are connected.

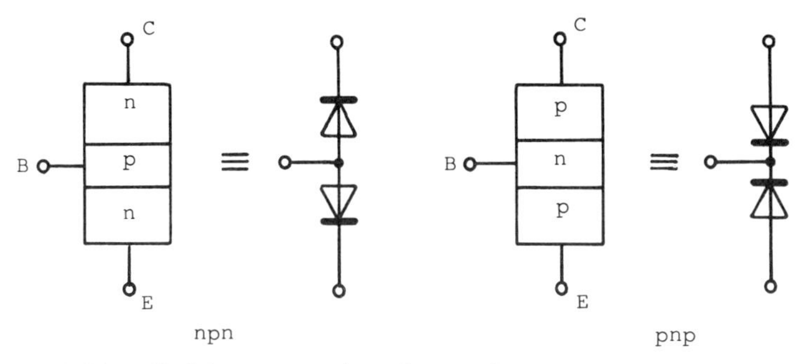

Fig. A1.6 'Two-diode' representation of a transistor

These three sets of readings are obtained regardless of which type of transistor (npn or pnp) is being tested. This enables to carry out a simple and quick test which in the majority of cases is all that is necessary to find out if a transistor is serviceable. If we take three pairs of readings with the ohmmeter between the three terminals, the six readings should produce

pair 1: 'high–high' (collector–emitter),
pair 2: 'high–low' (either base–emitter or base–collector),
pair 3: 'high–low' (either base–emitter or base–collector).

It is *not* necessary to know which terminal is which. If these readings are obtained, it is highly probable that the transistor is serviceable. If these readings are not obtained, the device is definitely faulty.

It is extremely unlikely that a servicing technician will not know, or be unable to find out, which type of transistor he is working with and which terminal is which; hence the above test will generally be sufficient for most cases. However, with a thorough understanding of the theory of semiconductors we can use the 'two-diode' concept to identify the three terminals, to establish whether the device is npn or pnp, and finally to get some indication of the current gain of the device. Figure A1.7 shows the polarities required to obtain the various readings on the two types of transistor.

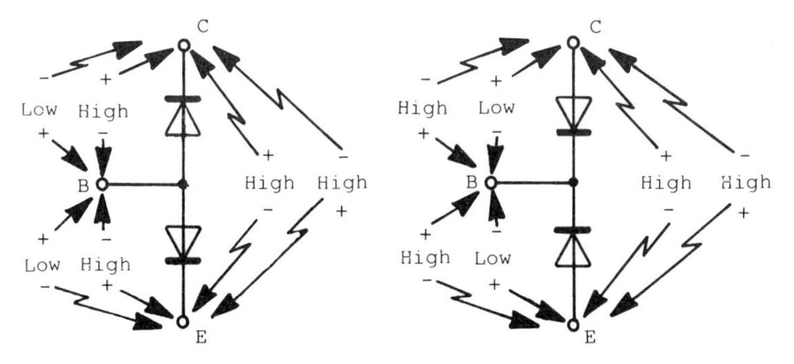

Fig. A1.7 Transistor resistances

1. To find the base terminal, simply find the pair of terminals between which two 'high' readings are obtained. The remaining terminal is the base.
2. To find out whether the transistor is npn or pnp, connect the positive lead of the ohmmeter (note: with the 'AVOMETER' and most multimeters the negative terminal is positive on the ohms range) to the base and connect the negative lead to one of the other terminals – it does not matter which. If a 'low' reading is obtained the transistor must be npn, and vice versa.
3. To determine which of the remaining terminals is the emitter and which the collector, measure the resistance between the two 'remaining' terminals – noting the resistance reading *and* the polarity connected to each terminal (use the highest range available on the ohmmeter). The resistance is highest between collector and emitter when biasing for normal operation is achieved, i.e. with positive to the collector for an npn transistor. This is shown in fig. A1.8.

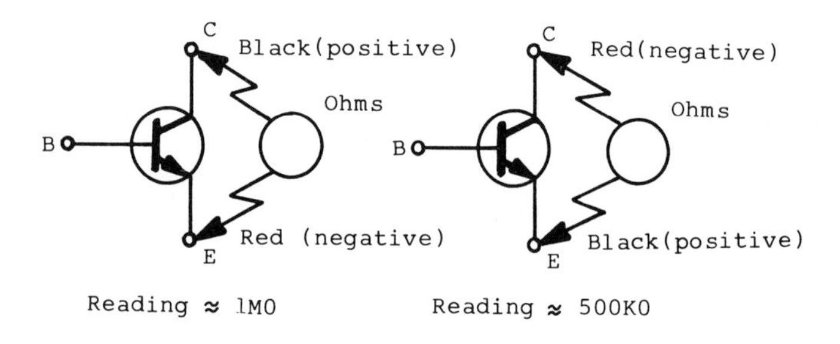

Fig. A1.8 Determination of collector and emitter terminals

4. To test for current gain, having carried out the other tests satisfactorily, connect the ohmmeter leads for correct operation – i.e. positive to the collector and negative to the emitter for npn. Using a suitable high resistance (a wet finger), connect the collector to the base (do *not* short them). A 'good' transistor will produce a significant decrease in the indicated resistance. If the collector and emitter are reversed, the reduction in resistance will not be as large. This test is not suitable for use on power transistors.
5. If possible, obtain silicon and germanium npn and pnp transistors and use them to carry out measurements to test the theory in steps 1 to 4.

Exercise A1.5 FET characteristics

1. Connect up the circuit of fig. A1.9, using a BFW10 JUGFET.
2. Set V_{GS} = 0 V and record the values of drain current for convenient values of V_{DS}.

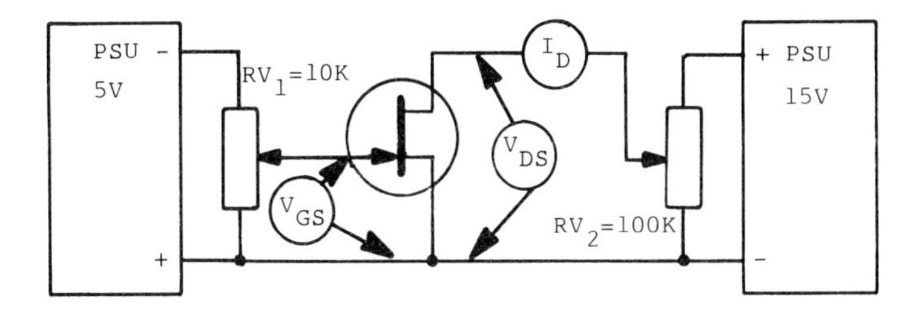

Fig. A1.9 Circuit for exercise A1.5

3. Repeat step 2 for values of $V_{GS} = -1$ V, -2 V, and -3 V.
4. Set $V_{DS} = 10$ V and record the drain current for V_{GS} from 0 V to -3 V in 0.5 V steps.
5. Plot the drain (output) characteristics and from them obtain

   a) g_m for $V_{DS} = 10$ V,
   b) r_{ds} for $V_{GS} = -2$ V.

6. Plot the mutual characteristics and obtain g_m for $V_{DS} = 10$ V.
7. Comment on any differences between the values obtained for g_m in steps 5 and 6.

Exercise A1.6 Basic d.c. regulator

1. Connect up the circuit of fig. A1.10, using an LM340-T5 regulator chip.
   Note: if the regulator is to be used as a power supply, it is best to mount it on a heat sink.
2. Measure the voltages at the input and output of the regulator using

   a) a d.c. voltmeter,
   b) a CRO.

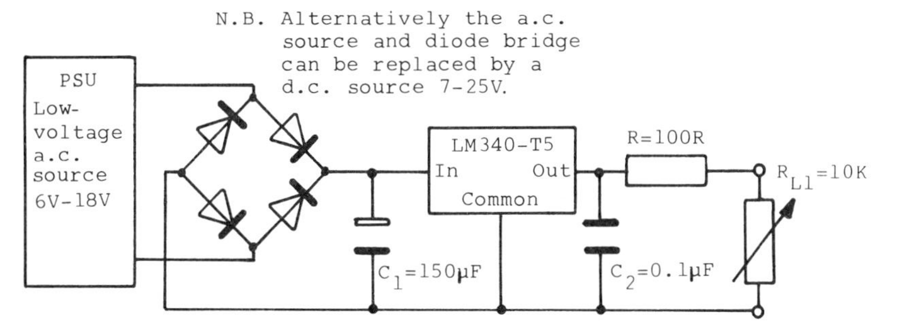

Fig. A1.10 Circuit for exercise A1.6

3. Use the 'a.c.' facility of the CRO to measure the 'ripple' at the input and output of the regulator.
4. Observe the effects on the output voltage of varying the load on the regulator using R_{L1}.
5. Comment on the readings obtained in steps 2, 3, and 4.

Exercise A1.7 Basic transistor bias

1. Connect up the circuit of fig. A1.11, using a BC109 transistor.
2. Using R_{V1}, set $V_{BE} = 0$ V and record the value of V_{CE}.
3. Increase V_{BE} slowly and observe the effects on V_{CE}. Record the minimum value of V_{CE} – this is $V_{CE(SAT)}$ – and compare this with the value given in the data sheets at the end of chapter 2.
4. For a range of values of V_{CE} between its maximum and minimum, record V_{CE} and hence calculate I_C (– ($V_{CC} - V_{CE}$)/R_C) and use this data to plot the load line for the circuit.
5. Adjust V_{BE} until $V_{CE} = 5$ V. Hold the transistor between finger and thumb and observe the effects of temperature on the collector voltage.

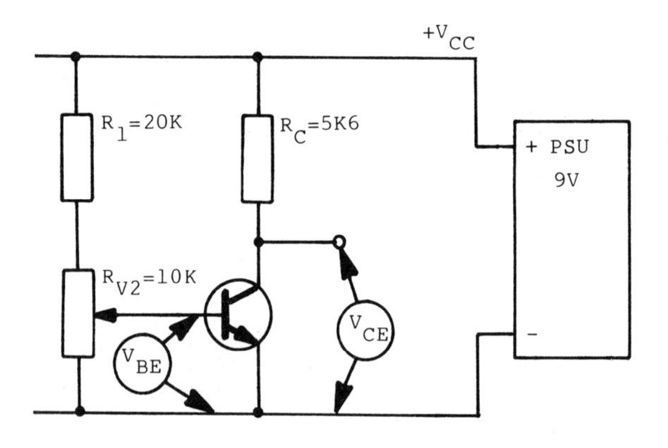

Fig. A1.11 Circuit for exercise A1.7

Exercise A1.8 Common-emitter amplifier

1. Connect up the circuit of fig. A1.12, using a BC109 transistor.
   Note: the input signal in fig. A1.12 is shown fed in to the amplifier via a potential-divider, R_{V3}. This is not necessary if the output of the signal generator can itself be reduced sufficiently, but this was not possible with the Feedback signal generator used in testing the circuits.
2. Measure the d.c. voltages V_{CC}, V_C, V_B, V_E, and V_{BE}.
3. Comment on any difference between V_{BE} and $V_B - V_E$.

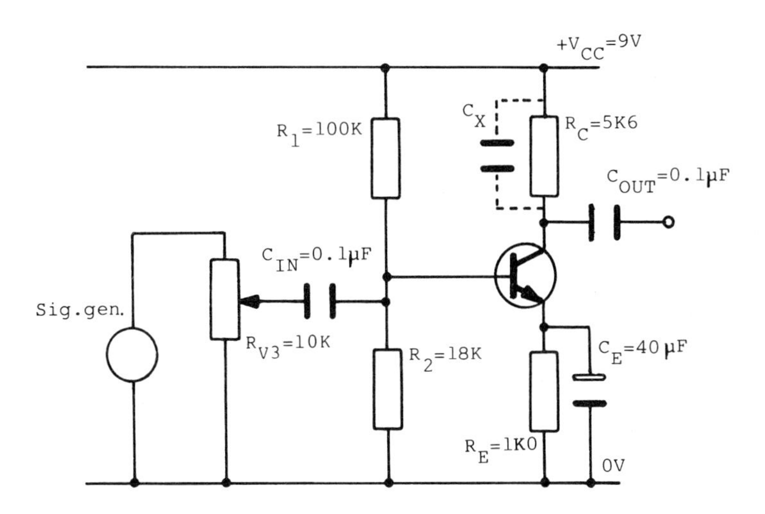

Fig. A1.12 Circuit for exercise A1.8

4. Inject an a.c. input signal at a frequency of 1 kHz, to give a peak-to-peak voltage of 20 mV at the base of the transistor, measured on the CRO. Observe both input and output waveforms simultaneously, using *both* channels of the CRO.

   Using the CRO time-base control, obtain the smallest possible number of complete cycles of the waveforms on the CRO screen. Calculate the frequency of the signal waveform. Does this agree with the value on the signal generator?

   Observe the phase shift between the input and output waveforms of the amplifier. Does it agree with the predicted phase shift?

   Measure the *peak-to-peak* voltage of both waveforms and calculate the gain of the amplifier:

$$A_V = \frac{V_{OUT(P-P)}}{V_{IN(P-P)}}$$

5. Increase the input signal to 200 mV (peak-to-peak on the CRO). Observing the output waveforms now shows that the output is no longer the same 'shape' as the input: it is badly distorted by 'clipping'. You should be able to explain what is happening by referring to the load line and characteristics.

   Using the CRO, measure the gain A_V. Can you explain the reason for any differences between this value of gain and the value obtained in step 4?

   Connect a multimeter on the a.c. voltage range across the output in parallel with the CRO. What effect does this have on the output waveform? Why?

Remove the multimeter. Disconnect the emitter bypass capacitor C_E. Calculate the new value of gain. Is it what you would expect? What has happened to the distortion? Explain why.

6. Reconnect the emitter bypass capacitor. For a range of frequencies from about 10 Hz to the maximum for your signal generator and an input signal voltage of 20 mV peak-to-peak, measure the output voltage (peak-to-peak) on the CRO. Calculate the gain and plot gain against frequency to determine the '3 dB points' (the points where the gain has fallen to 0.707 times the 'mid-band' value); hence find the amplifier bandwidth.

   It will not be necessary to take large numbers of readings in the mid-frequency ranges. Over the range of frequencies where the gain is changing, take more readings. Since the output voltage of the signal generator will change as the frequency (and hence the loading) is changed, it will probably be necessary periodically to adjust the input signal level. In practice, it is most convenient to keep the input signal, as measured on the CRO, constant – say by keeping it to 4 cm peak-to-peak on the 5 mV/cm setting.

   Some older signal generators do not have a sufficiently high frequency range to achieve the fall-off in gain at the high-frequency end of the amplifier range. In this case, a capacitor $C_X = 0.1\ \mu F$ across R_C will generally achieve a sufficient fall-off in gain to demonstrate the effects.

7. Measure the phase-shift between the input and output signals at the '3 dB points' found in step 6.
8. Plot the gain/frequency characteristic and determine the bandwidth of the amplifier.
9. If time permits, repeat steps 6, 7, and 8 with the emitter bypass capacitor disconnected. It may be necessary to increase the input signal to calculate the gain in this case.
10. Switch power off. Measure and record the resistance of each resistor in the circuit. For each resistor take two readings – with the multimeter both ways round.
11. Explain why the two readings across each resistor in step 10 give different readings. (Hint: do any parallel paths have diodes in them?)

Exercise A1.9 Input and output impedances

The input and output impedances of amplifiers are very difficult quantities to measure accurately. The circuits necessary to make these highly technical measurements are only available in the most sophisticated design and test establishments.*

Some indication of the values of input and output impedance can be found by considering the amplifier to have a purely resistive load and

* Refer to *Transistor circuit design* by Texas Instruments Inc. (McGraw-Hill, 1963) for typical circuits and methods.

neglecting any inter-electrode or stray capacitance. The impedances can now be considered to be purely resistive. However, even with these simplifications, the necessary measurements are not easy to make with a high degree of accuracy – mainly due to the difficulty of measuring small a.c. currents.

The circuit of fig. A1.13 can be used to obtain typical values for these two parameters.

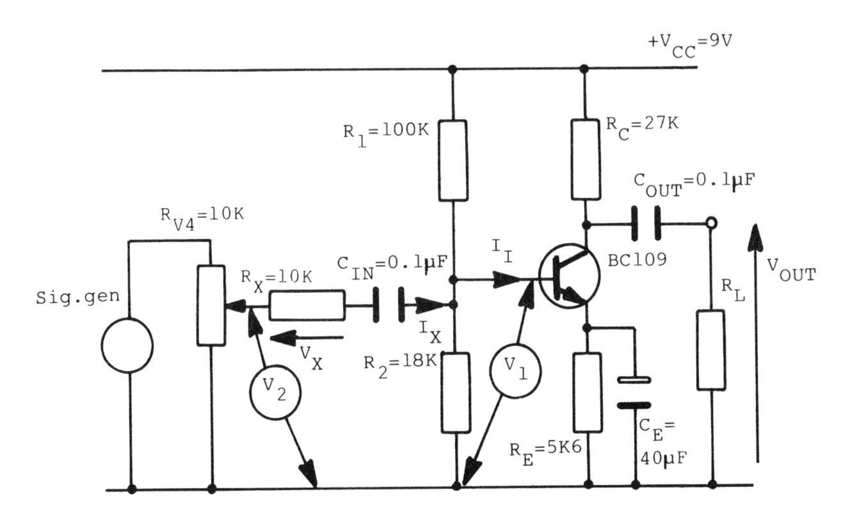

Fig. A1.13 Circuit for exercise A1.9

Input impedance

This can be found quite simply using Ohm's law and is given by V_1/I_I. Note that for a.c. purposes $V_1 = V_{BE}$, since R_E is effectively short-circuited by C_E over the frequencies being considered. As stated previously, the difficulty is in measuring I_I. Provided R_1 and R_2 are large compared to the input impedance of the transistor, a reasonable approximation can be obtained by measuring the voltage across R_X and calculating

$$I_I \approx I_X = V_X/R_X = (V_2 - V_1) \times 0.707/2R_X$$

(Note: 0.707/2 converts peak-to-peak values to r.m.s.).

1. Set the signal-generator frequency to 10 kHz. Connect the Y_1 channel of the CRO to measure V_1.
2. Adjust the potentiometer until the input signal is 30 mV (6 cm on the 5 mV/cm setting).
3. Now connect the Y_1 channel to measure V_2 and calculate

$$I_X = \frac{V_2 - V_1}{R_X} \times \frac{0.707}{2}$$

Hence calculate $R_{IN} = V_1/I_X$.

Compare the value obtained with the value quoted for the BC109 in the manufacturer's data sheets at the end of chapter 2.

Output impedance

This is the most difficult parameter to obtain using simple measurement techniques. The method below provides a reasonable value for the parameter. It uses the maximum-power-transfer theorem.

It can be shown that maximum power is transferred between a source and a load when the load and source impedances are equal. If we plot the power delivered against the effective load on the transistor, we can deduce the output impedance if we know or can calculate the external load resistance.

Figure A1.14 shows the effective equivalent circuit of fig. A1.13. If various values of R_L are used and the output voltage V_{OUT} is measured, the power output, $P_{OUT} = V_{OUT}^2/R_P$, can be calculated. If the output power is plotted against equivalent load resistance R_P, an approximate value for R_{OUT} can be found.

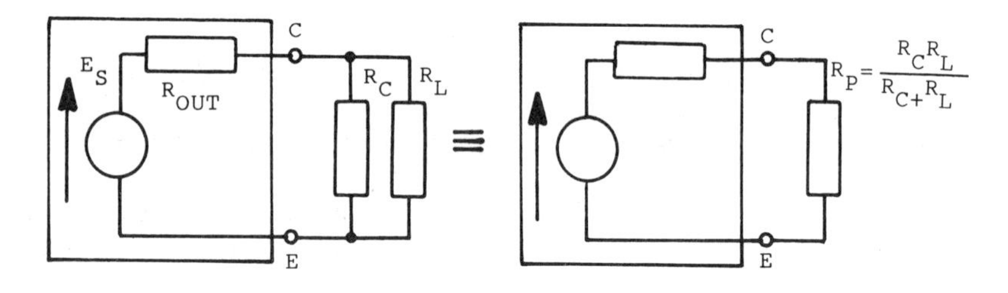

Fig. A1.14 Equivalent output circuit of fig. A1.13

1. Set the signal-generator frequency to 10 kHz and the input voltage V_1 to 30 mV peak-to-peak. For a range of values of R_L, calculate R_P, measure V_{OUT}, and hence calculate P_{OUT}. R_{OUT} can then be determined by plotting a graph of R_P against P_{OUT}. The values for R_L should be chosen so that it is probable that the equivalent resistance R_P will cover the expected range of possible values for R_{OUT}. Also, R_C must be larger than the value expected for R_{OUT}.

It is *essential* that the greatest possible care is taken over all measurements of voltage for this particular exercise. The data in Table A1.1 for a BC109 was obtained using the circuit of fig. A1.13. The values of R_L were set on a potentiometer before the voltage measurements were recorded.

Using the values in Table A1.1 an output resistance of about 20 kΩ was obtained.

Table A1.1

R_L (kΩ)	$R_P = \dfrac{R_C \times R_L}{R_C + R_L}$ (kΩ)	V_{OUT} (V)	$P_{OUT} = \dfrac{V_{OUT}^2}{R_P}$ (μW)
5.5	4.6	0.72	15
10	7.3	1.16	23
20	11.5	1.8	35
30	14.2	2.2	43
40	16.1	2.5	49
50	17.5	2.65	50
60	18.6	2.85	55
70	19.5	2.95	56
80	20.1	3.05	58
100	21.25	3.0	53
110	21.6	3.0	52

Exercise A1.10 Two-stage common-emitter and common-source amplifiers

1. Connect up the circuit of fig. A1.15.
2. Measure and record the d.c. collector, base, emitter, and base–emitter voltages of the transistors.
3. Connect the signal generator to the input terminals and use the CRO to observe the input and output waveforms of the amplifier.
4. Set the frequency output of the signal generator to 1 kHz. Adjust the amplitude of the output until the output waveform just starts to distort.

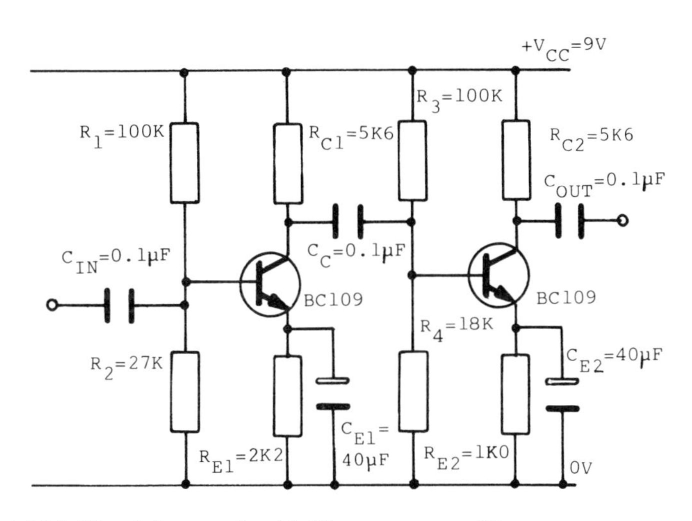

Fig. A1.15 Circuit for exercise A1.10 – common-emitter

Observe the phase shift of the amplifier. Is it what you would expect?

5. Reduce the input signal to a convenient value of about half that which just produced distortion in step 4. Measure and record the amplifier gain at 1 kHz.
6. Find the bandwidth of the amplifier, using the method described in step 6 of exercise A1.8.
7. Connect up the circuit of fig. A1.16 and repeat steps 1 to 6.
8. Is there any difference in the bandwidth between the two amplifiers of figs A1.15 and A1.16? If so can you explain why?

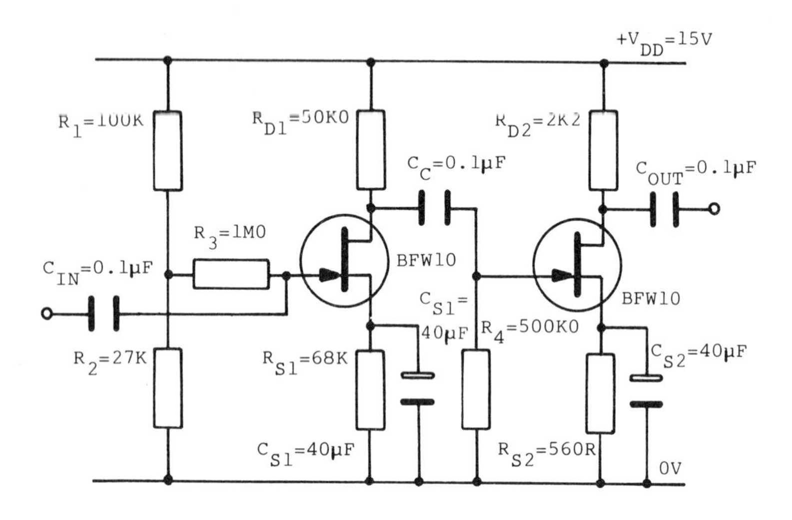

Fig. A1.16 Circuit for exercise A1.10 – common-source

Exercise A1.11 Tuned-load amplifier

1. Connect up the circuit of fig. A1.17.
2. Calculate the resonant frequency of the tuned circuit.
3. Adjust the input signal to 20 mV peak-to-peak as measured on the Y_1 channel of the CRO.

   Set the signal-generator frequency to the value calculated in step 2. 'Swing' the frequency slowly around to find the exact frequency of resonance. Measure the output voltage at resonance on Y_2.

   Calculate the voltage gain at resonance.

   Determine the amplifier gain at a range of frequencies on either side of the circuit resonant frequency.
4. Plot the gain against frequency and obtain the bandwidth of the amplifier.

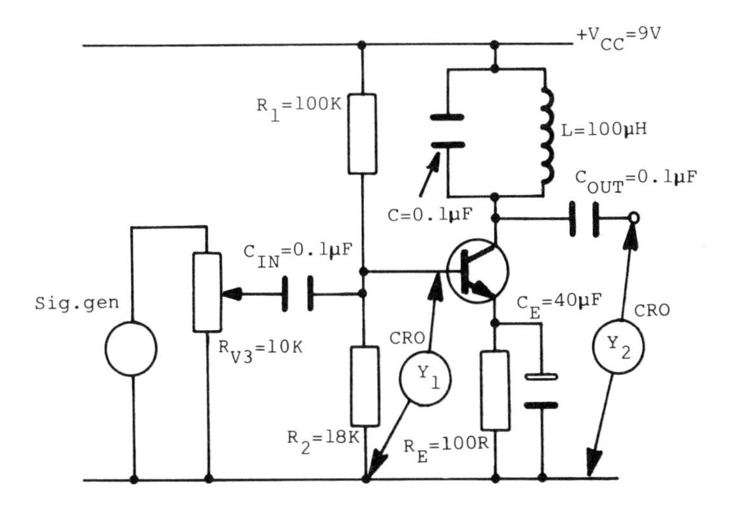

Fig. A1.17 Circuit for exercise A1.11

Answers

Answers are provided to all the numerical questions and to some of the multi-choice and short-answer questions. In the case of answers requiring data to be obtained from graphical work it should be remembered that slight differences in readings are quite normal, resulting in minor differences in answers.

Chapter 1

1.1 The number of electrons in the outer shell or band
1.2 Electrons **1.3** Cathode to anode **1.4** (d) **1.5** (f)
1.6 (e) **1.7** (e) **1.8** (b) **1.9** Electrons, heat **1.10** (a)–(c) and (e) 5A, (d) 0 **1.11** 212 mA **1.13** (a) 286 Ω, say 0.25 W; (b) 9 mA
1.14 (a) 414 Ω, say 1 W; (b) 40 mW **1.15** (a) 74 mA, (b) 42 mA
1.16 (a) 1.3 W, (b) 500 mW

Chapter 2

2.1 (a) 60, (b) 5.4 kΩ, (c) 400 Ω **2.2** (a) 0.99, (b) 200 kΩ, (c) 28 Ω
2.3 (b) **2.4** Low **2.7** Common-emitter

Chapter 3

3.1 580 kΩ (assuming $V_{BE} = 0.2$ V) **3.2** (a) 38 kΩ, (b) 179 mW, (c) 95, (d) 38, (e) 3610 **3.3** (a)–(e) and (h) 0 V; (f), (g), and (i) 10 V
3.4 (a) Q = (5.1, 30.2); (i) 313 kΩ, (ii) 4.9 V, (iii) 30.2 mA, (iv) 154 mW, (v) 146 mW, (vi) $\approx I_{CQ}V_{CQ} = 300$ mW; (b)(i) 1075, (ii) 43.5 dB, (iii) 51.9 dB **3.5** (a) 840 kΩ; (b)(i) 53.5 dB, (ii) 47.6 dB, (iii) 50.5 dB **3.6** (b) $R_B = 135$ kΩ, $R_C = 513$ Ω; (c)(i) 35.9 dB, (ii) 43.3 dB, (iii) 39.6 dB **3.7** (a) 293 Ω, (b) 29.7 W, (c) 28.5, (d) 15 dB **3.8** (a) Using $I_{BQ} = 150\ \mu$A, $A_I = 235$; (b) 67
3.9 (a) Q = (4.6, 3.65) and $I_{BQ} = 50\ \mu$A, (c) 53 **3.10** (a) Q = (4.9, 4.5); (b)(i) 470, (ii) 135 **3.11** (a) 476; (c)(i) 53.5, (ii) 12.7, (iii) 718 (use R_P not R_C) **3.12** Convert current to voltage **3.13** (f)
3.14 Higher **3.21** Increase value of coupling and decoupling capacitors

Chapter 4

4.1 16.9 **4.2** 18.3 **4.3** (b)(i) 3.75, (ii) 1.33 V
4.4 (a)(i) 3.55 mS, (ii) 20 kΩ; (b) 4.33 mS **4.5** (a) $R_D = 8.12$ kΩ, $R_S = 211$ Ω; (b) 8.75 **4.6** (b) 4.5 (note: you must interpolate between $V_{GS} = -1$ V and $V_{GS} = -2$ V curves); (c) 16.3 kΩ (note: $V_{GS} =$

$-1.5\text{ V} = V_G - V_S = V_G - I_{DQ}R_S$, $\therefore V_G = V_{GS} - V_S = -1.5\text{ V} - 2.18\text{ V} = -0.68\text{ V}$, $\therefore R_2 = V_G/I_2 = 0.68\text{ V}/41.6\ \mu\text{A}$)

Chapter 6
6.1 30 dB, 44 dB, 0 dB, – 3 dB, 56.5 dB, 60 dB **6.2** 25 mW
6.3 40.4 dB, 33.4 dB **6.4** (b) 251.2 kHz – 63 Hz, ≈ 251.2 kHz; (c) $E_N = 2.84\ \mu$V (assuming ambient temperature of 20°C)

Chapter 7
7.1 0.0399 **7.2** 39.9 **7.3** (a) 824 Ω, (b) 99.99
7.4 (a) 158.5 kHz – 79 Hz, ≈ 158.5 kHz; (b) ≈ 2 MHz
7.5 (a) 101, (b) 100, (c) 49.5, (d) 0.323%

Chapter 10
10.1 0.18 A **10.2** (a) 30°, (b) 8.4 A **10.3** (a) 60°; (b) 6.84 W; (c) Power dissipation of SCR = $I_{MAX} \times 1$ V = 0.1125 W, say ¼ W rating **10.4** (c) **10.5** All triacs **10.6** 25 A, 25 W

Chapter 11
11.1 8 **11.2** 9.2 **11.3** (a) R_1 = 100 kΩ, R_2 = 20 MΩ; (b) 0.004 995 **11.4** $R_1 \approx R_{OUT} \approx$ 2 kΩ, R_2 = 118 kΩ

Chapter 12
12.1 (a) 200 kHz, (b) 1:2.33 **12.2** (a) ≈ 0.5 μs, (b) ≈ 1.5 μs, (c) 10 μs **12.3** 25.99 kHz **12.4** (a) 25.33 mH, (b) 44.7 kHz
12.5 (a) 43.42 kHz, (b) 1:2.5 **12.6** R_2 = 507 Ω, R_3 = 2.53 kΩ

Chapter 13
13.1 (a)(i) 8.28 kΩ, (ii) 0.1035 Ω; (b)(i) 77.24 kΩ, (ii) 263.91 kΩ; (c) 0.1035 maximum **13.2** (a) 6.07 V; (b)(i) 3.18 V, (ii) 5.565 V
13.3 (a) 26.1 μA, (b) 25.82 μA, (c) 4.6 V, (d) 4.7 V **13.4** (a) 1.56 V; (b)(i) 1.46 V, (ii) 6.4%; (c)(i) 1.53 V, (ii) 1.96% **13.5** (a) 147.6 Ω, (b) 148.05 Ω and 147.16 Ω, (c) None **13.6** 350 Ω **13.7** (a) 55 mV if a square wave, 110 mV if a pulse; (b) 40 kHz **13.8** (a) ≈ 62 mV, (b) 3.18 V, (c) 51.4, (d) 500 kHz, (e) 54°

References

As mentioned in the preface, in teaching one invariably consults many sources when preparing material for teaching notes – e.g. books, technical journals, data books, etc. – and once again I wish to acknowledge my debt to those long-forgotten sources. Some of the particular sources I have found useful during my teaching career and which have contributed to the material in this book are listed below and can be referred to by those wishing to obtain a deeper understanding or wider knowledge than is necessary for those studying for TEC awards.

1. *Mullard minibook* series

2. *A programmed book on semiconductor devices* (1970)

Both the above books are Mullard Educational Service publications. Unfortunately these excellent little books are no longer in print.

3. *Transistor circuit design*, Texas Instruments Inc. (McGraw-Hill, 1963)

4. *Understanding solid-state electronics*, Texas Instruments Learning Centre, 1972. (This is now published by Radio Shack and sold by Tandy in the UK.)

5. *Transistor manual* (7th edition), General Electric Company, USA (1964)

6. *SCR manual* (6th edition), General Electric Company, USA (1979)

7. *Circuits, devices, and systems*, R. J. Smith, (Wiley, 1976)

8. *Transistor pocket book*, R. G. Hibberd (Newnes, 1976)

9. *Electronics pocket book* (4th edition), edited by E. A. Parr (Newnes, 1981)

10. *Half-way elements: the technology of metalloids*, G. Chedd (Aldus, 1969)

11. *The cathode-ray oscilloscope and its use*, G. N. Patchett (Norman Price, 1976)

Key to TEC objectives

General objective	TEC units and relevant sections or chapters of the book		
	U81/747	U81/743	U81/742
A1		2.6	
A2		4.2 to 4.5	
A3		4.6 to 4.9	
B4		2.4, ch. 3, ch. 8, 9.2, 9.3, 9.7, appendix	
B5		9.2 to 9.6, 11.2, 11.3	
C6		Chapter 6	
D7		Chapter 7	
E8		12.1 to 12.9	
E9		12.10 to 12.13	
F10		Chapter 11	
G8			Appendix
G9	1.1 to 1.5		
G11		Chapter 5	
H9			Chapter 10
H10	2.1 to 2.5		
I11	13.2 to 13.7, appendix		

Index